£2.50

ZZAAP!

ZZAAP!

Taming ESD, RFI, and EMI

M. Bruce Corp

ACADEMIC PRESS, INC.
Harcourt Brace Jovanovich, Publishers
San Diego   New York   Boston
London   Sydney   Tokyo   Toronto

This book is printed on acid-free paper. ∞

Copyright © 1990 by Academic Press, Inc.
All Rights Reserved.
No part of this publication may be reproduced or transmitted in any form or by any means, electronic or mechanical, including photocopy, recording, or any information storage and retrieval system, without permission in writing from the publisher.

Academic Press, Inc.
San Diego, California 92101

United Kingdom Edition published by
Academic Press Limited
24–28 Oval Road, London NW1 7DX

Library of Congress Cataloging-in-Publication Data

Corp, M. Bruce.
    Zzaap! : Taming ESD, RFI, and EMI / M. Bruce Corp.
      p. cm.
    Includes bibliographical references.
    ISBN 0-12-189930-6 (alk. paper)
    1. Microprocessors. 2. Electronic circuit design.
3. Electromagnetic compatibility. I. Title.
TK7895.M5C67 1990
621.381–dc20   90-31990
                                                                             CIP
                                                                              r90

Printed in the United States of America
90 91 92 93   9 8 7 6 5 4 3 2 1

Contents

Preface vii

Acknowledgments xi

Introduction 1

CHAPTER 1
The "Why and How" of Computer Crashes 7

CHAPTER 2
AC Power Variations 33

CHAPTER 3
Surge Suppressors and Noise Filters 59

CHAPTER 4
Grounding Requirements and Lightning 81

CHAPTER 5
Noise Susceptibility and Emission 103

CHAPTER 6
Other Reliability Factors 121

CHAPTER 7
Why "Burn-In"? 147

CHAPTER 8
Environmental Effects and ESD 157

CHAPTER 9
Thermal Shock 173

CHAPTER 10
Preventive Design Techniques 185

CHAPTER 11
Printed-Circuit-Board Design and Layout 195

CHAPTER 12
Overall System Design 205

APPENDIX
Other Sources of Information on This Subject 221

Glossary 223

Index 229

Preface

For the past several years I have been actively involved in the design and integration of computers, solving the "after-the-fact" problems arising from the marriage of computer systems to unprepared installation sites and facilities. Frustration runs rampant in a situation in which a computer is integrated into a facility by an inexperienced software-house-turned-integrator. Usually, little or no site preparation is done because it is not thought to be needed. And the peripherals included with the system are usually chosen with only one thought in mind—cost.

This book is about the "why and how" of many computer crashes and their prevention. It is a source of priceless information useful to anyone who is involved with the design, debugging, or development of microprocessor-based equipment. *ZZAAP! Taming ESD, RFI, and EMI* covers system crashes that are caused by reasons other than software bugs. Equipped with this book, you will be able to tell the difference between software bugs and a "ZZAAP!" Electrical noise types and their sources, along with the details of noise susceptibility and noise emission, are thoroughly explored in this book.

ZZAAP! contains information normally unavailable to anyone not already "in the know." Usually, the people who have specialized expertise in the area of AC power, grounding, radio-frequency interference (RFI), electromagnetic interference (EMI), electrostatic discharge (ESD), noise susceptibility, emissions, and so on, do not talk about it. They feel this information is proprietary and are in no hurry to make it widely known.

Expertise in this area comes from many years of experience in both analog and digital design—and from having already made the mistakes. It comes also from actual research into the causes of the various problems discussed in this book. Now, I have put all this into one comprehensive easy-reading source . . . *ZZAAP!*. I have included data in this book from both practical experience and research, as well as reference material from other sources. Countless "lessons learned" are available to the reader. Examples are accompanied by actual data gathered at the scene and completely analyzed in this book.

The results from these "lessons learned" are applied as design rules for microprocessor systems in a practical approach to preventive design techniques. Coverage of this subject will convince you that unless the original design incorporates good grounding and filtering design techniques from the beginning, electromagnetic emissions and/or susceptibility to radiated and conducted interference can be inadvertently "built-in."

This book features a comprehensive, engineering-level description of the power, grounding, and environmental bugaboos of microcomputer system design. *ZZAAP!* is rich in theory, but with the equations presented in a practical person-to-person way. I hope you will find this is not just another dry textbook. Practical application of theory is demonstrated through actual case histories. You will have the opportunity to be a part of investigative research into such failure causes as:

- the effects of voltage variations (spikes, sags, "glitches," power factor, RF noise, etc.)
- ground noise sources and ground gradients caused by ESD, lightning, thunderstorms, soil composition, etc.
- ambient temperature effects on performance (with methods for controlling or minimizing its effects)
- thermal shock and its effects on systems
- "infant mortality," with a very scientific response to "why burn-in?"
- reliability factors beyond the control of the manufacturer

You may find yourself wondering why I included coverage of such things as ambient temperature effects, thermal shock, and burn-in in a book on ESD, RFI, and EMI. The reason is simple: this book is actually a treatise on *computer system reliability from a designer's viewpoint,* and "how to design it in from the ground up" is the central theme. Reliability covers more territory than just outside environmental effects on a system, so I thought it worthwhile to devote space to these other concerns.

I have organized the book so you can go directly to a subject's area of coverage by following the table of contents. The glossary at the back of the book will help you to understand some of the terms you may not immediately recognize, giving you some of the "buzz-words" used in this field.

Please read the introduction—it contains some information not found anywhere else. It will give you the "theme" for the entire book. I have tried to present this otherwise dry material in a light, enjoyable-reading way, yet still provide you with the real facts and data.

Good luck, and happy reading!

Acknowledgments

I would like to thank the following people for their very helpful support during the production of this book. First and foremost, I wish to thank my wife, Viola, for her encouragement, her loving and gracious patience, and for bringing me meals during my long hours of beating on the keyboard of my Macintosh computer. A very special loving thanks to my Aunt Francis and Uncle Earnest, whose inexhaustible support so greatly influenced my early married life. Special thanks to Ken Motoh (the best technician an engineer ever had), for his standing-by friendship through all the heartache of our learning curve. Thanks to Mr. Gail B. Mathas for his kind help when it was most needed. It was gratefully appreciated. A special thanks to Mr. John Howard of Radiation Technology in Morgan Hill, California, for his friendship, understanding, and technical knowledge. Also, I wish to thank John for his unselfish efforts in trying to further the individual pursuits of his business associates and personal acquaintances. Thanks to the well-known consultant in this field, Mr. J. F. Kalbach, for getting me interested in this field of expertise to begin with. Also, thanks for helping me to realize that the field was a "natural" for my particular background.

Introduction

Have you ever found yourself in the following very frustrating position? Let's assume you (or someone you know) are in the middle of a program—and suddenly the computer stops responding to the keyboard. It just sits there, apparently dead to the world, the last screen still displayed, but paying no attention to anything you do. In final desperation, after trying everything else you can think of, you give up and hit the "reset" button . . . while realizing you have lost the whole last 2 hours' work.

Sound familiar?

On the other hand, sometimes less destructively . . . but still a problem—one of the following happens:

1. Bright "flashes" appear across the screen intermittently.
2. Disk error messages from the system start appearing while the disk drive is running.
3. For no apparent reason, the system screen fills with "garbage."

Or suppose a machine that is controlled by an embedded microprocessor doesn't stop when it is supposed to and over a thousand pieces are ruined before the operator notices there is something wrong with it! The machine has to be powered-down and totally reset to get it operating again. Since this was an embedded processor, the microprocessor's software was contained in read-only-memory (ROM) and therefore could not have been erased or changed. There were plenty of sensors provided to sense this kind of failure, but it must have simply been ignored by the computer. Why were the sensors ignored? Probably because the processor was off in "Never-Never Land."

All the foregoing scenarios have one common denominator: the question of whether it was a software crash or a hardware failure. The truth is, it was probably neither a hardware failure *nor* a software-induced crash. More than likely, the computer became a victim of the villain of this book—a scoundrel we shall call "Zzaap!"(Fig. 1).

If you have not experienced one of these episodes yourself (and you spend much time around computers), you should consider yourself very lucky, indeed. Most probably, at least one of these *has* happened to you, and you, too, have been bitten by one of the most ubiquitous problems associated with computers . . . and referred to by such names as "lock-up," "freeze," "hang," or "crash."

This book will help you to sort out a "zzaap" from a software-induced system crash. It will also give you the insight necessary to design out these problems. The events we have described happen entirely too often. Anyone who has been in the situation described knows the feeling of utter helplessness

Figure 1 The Scoundrel "Zzaap!"—the villain of this book!

and sense of loss that comes when it does happen. But the point is, *it doesn't have to happen to anyone.*

Yet such situations bring up the following questions:

- Was this a software system crash, a hardware failure, or a power or ground transient?
- Is it enough to explain to the victim that the document they were working on need not have been lost if they had saved to disk often enough?
- Is saving the document every 15 minutes the only weapon we have against unforeseen crashes?
- Shouldn't someone have "tried a little harder" to prevent this kind of problem?
- How can we learn what causes this problem and prevent it, before it happens (rather than trying to repair the results)?

The subject material contained in this book may, at first glance, seem very trivial. But millions of dollars have been lost by companies who *thought* it was trivial, and didn't pay it the amount of attention it deserves. Millions more have been spent on finding out (after the fact) "What happened?" and "What could we have done to avoid it?"

ZZAAP! Taming ESD, RFI, and EMI brings it all together—everything you ever wanted to know about the why and how of computer crashes and their prevention, from design to implementation and installation. As you will see, anytime a computer randomly "freezes," "locks up," or stops responding to keyboard inputs, it has likely become a victim of a "zzaap!"

What is a "zzaap?" It is sometimes very hard to separate the *cause* of an event (or the event, itself) from the *effect*. But for the purposes of this book, a "zzaap" is herein defined as the *effect* of an electrical impulse, sometimes called a transient (which is short for "transient impulse"), on a computer system. What kind of a transient impulse could cause the described symptoms in our scenario? Where did it come from? How did it affect the system, so that the particular symptom described was exhibited? A "zzaap" can be the result of any one of the following:

1. *Electrostatic discharge* (ESD) (more commonly called "static electricity"). ESD is the high-voltage "spark" caused by a static charge such as that resulting when you walk across a rug and then touch a metal object.
2. *Radio frequency interference,* (RFI). RFI, as defined here, is interference caused by radio-frequency energy manifested where it does not belong. RFI can be carried through air or by wire as a "stray" radio signal.
3. *Electromagnetic interference* (EMI). The trend lately has been to use the terms "EMI" and "RFI" interchangeably. Actually, EMI is a class of interference caused mainly by a varying magnetic field at the lower frequencies, rather than radio-frequency energy—and usually emanates from an electromechanical device. It can be propagated in two ways: either directly through the air (called radiated EMI) or by the varying magnetic lines of force cutting a conductor, causing an unwanted signal in that conductor (which we call conducted EMI).
4. *Sags.* A sag is a corresponding decrease in input alternating-current (AC) voltage
5. *Surges.* A surge is a sudden increase in the input AC voltage.
6. *Noise.* Noise is a term loosely used to describe any extraneous electrical signals that appear where they do not belong. "Noise" may even be loosely applied to the three causes (ESD, RFI, EMI) previously mentioned.

Welcome to the world of the computer power and ground *noise*. "Noise," in computerese, doesn't mean the kind your son or brother plays on his stereo, or the kind that comes from under the hood of your car when it is low on oil. The noise we speak of here is electrical impulses carried through the air, or induced into wires . . . And not just *any* electrical impulses, but a very special kind—*random, incoherent* electrical impulses.

If these impulses travel on wires, we say they are *conducted*. If they are carried through the air, we say they are *radiated*. In either case, they are unwelcome and must be dealt with. And that's what this book is all about!

Included in *ZZAAP!* are several other topics such as AC power problems and their remedies, lightning, "burn-in," thermal shock, and several other seemingly unrelated subjects. But as you'll see, these topics will provide very revealing insights into the mechanisms behind every type of failure mode known. You will be led through these and other subjects in a conversational manner, to provide a background of usable knowledge rather than to give you a set of unyielding rules to follow.

So sit back, relax, and enjoy a cruise through "The dark side of the Force!"

CHAPTER 1

The "Why and How" of Computer Crashes

Suppose you were in the middle of a program on your favorite computer (if there *is* such a thing), and it suddenly refuses to pay any attention to you, or the keyboard, the mouse, or whatever (i.e., the computer has "crashed"). A question immediately arises as to whether it was a software crash or a hardware failure. Actually, it was probably neither—it was most likely off in Never-Never Land as a result of a "zzaap."

In the Introduction, a "zzaap" was defined as an incident or event that is the *cause* of all the various effects that form the subject of this entire book. We will characterize a "zzaap" here as being the conspicuous evidence of a random *electrical impulse* (sometimes called a *transient,* or *transient impulse*) on a computer system.

Where does a "zzaap" come from, and how does it affect a computer system? Welcome to the wild, weird, wonderous world of the computer-equivalent of a nervous breakdown!

The event herein called a "zzaap" can be the manifestation of any of several forms of electrical disturbances. Most of these disturbances can be lumped under the general term "noise" to loosely describe any extraneous electrical signals that interfere with computers.

A "zzaap" can even be the result of the occurrence of an electrostatic discharge (ESD), more commonly called a *static spark*— the high voltage spark caused by a static charge, such as the one that results when you walk across a rug and then touch a metal object. A "zzaap" can also be the result of radio-frequency interference (RFI) caused by "stray" radio-frequency energy and manifested as an unwanted signal. RFI can be carried either by air (*radiated RFI*), or by wire (*conducted RFI*). Or the cause of the zzaap might even be electromagnetic interference (EMI). Contrary to the mistaken use of the terms "EMI" and "RFI" interchangeably, the definition used herein for EMI will be *the class of interference caused mainly by a varying magnetic field*. EMI can be generated by an electromechanical device or lightning. Like RFI, EMI can also be propagated through the air, or by wire. Last, but not least, the cause of a zzaap could be a surge or sag on the power line. A surge is a sudden increase in the input AC voltage. A sag, on the other hand, is a corresponding decrease in the input AC voltage.

Does this sound like a repetition of something you read before (e.g., in the Introduction)? Well it may be, but for a very good reason—it is very important, and should be remembered. We shall now explore each of these a little further . . . .

Electrostatic Discharge

As stated earlier, an electrostatic discharge (ESD) event is the "static spark" caused when a charged body approaches a conductive or oppositely charged object. If the object is conductive, it will provide a path for the ESD.

ESD (Fig. 1.1) is very much like miniature lightning. The only difference is the much reduced distances between the objects, which lowers the voltages and currents involved in the discharge. Even if the other object involved is not oppositely charged originally, it acts exactly as if the charged body were one plate of a capacitor and the opposing object were the other plate. The capacitance between the two plates causes the opposing object to take on an opposite polarity charge. As the charged body approaches, the air in between them then becomes ionized, making it even more conductive. As the distance between them further decreases, the voltage is finally able to

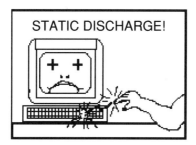

Figure 1.1 Electrostatic discharge. Electrostatic discharge can not only cause a system to "freeze," but can actually "blow out" some of the integrated circuits (ICs) in the circuit that happens to take a "hit." Proper grounding in the physical design can eliminate failures caused in this manner.

force current to flow to the other "plate" through the ionized air. This is the "flashover" point. At this instant, current flows, heating the air to incandescence and decreasing the level of charge differential between the two until the current stops. (Actually, this account is slightly oversimplified, but will be explained more fully later, with a complete explanation of the mechanism involved.)

The actions described in this simplified example can cause other, even more destructive results inside the computer. One such possibility is the high-voltage breakdown of any ICs unlucky enough to be in a path to the nearest ground. Another might be changes in the bit structure of data contained in memory—a phenomenon known as "picking a bit" (if the change in the bit is from a "0" to a "1") or conversely, "dropping a bit" if the change in the bit is from a "1" to a "0". . . . one of the most common causes of computer errors.

An electrostatic discharge doesn't even have to pass *through* the computer—it might simply take place between the charged body and a nearby ground. But the resulting impulse *induced* into nearby computer logic can do the trick. The amount of current in the static spark described here may be very low, but it takes place through a very high impedance (the air), with a very high voltage drop. At the instant of "flashover," an ESD spark causes very high levels of radiated RFI[1]—which in turn

---

1. Note that in this case, it is a high-voltage spark that causes the RFI rather than high currents, which can cause EMI.

can cause new problems not even related to the "static spark" itself, as we shall see.

Radio-Frequency Interference

A "zzaap" can be the result of radio-frequency interface (RFI) from a spark, as in the example above, or from a nearby radio or radar transmitter. In the foregoing discussion of electrostatic discharge, most of the noise frequencies generated by the static spark fall into a range of frequencies we know as the *radio range* (hence the name RFI). The fact that they are radio frequencies means that they can be either radiated through air, or conducted by wires. It also means they are frequencies above 50 kHz or so.

Extensive research has proved that very high intensity electrical noise with extremely steep rise-times are the most destructive to reliable computer operations. The exact mechanism by which each electrical impulse might affect a particular computer is a very complex thing to pin down, due to the tremendous number of variables involved. But we *can* list the variables (or most of them). Some of the variables involved are:

1. The frequency of the interfering signal. (Because the higher the frequency, the less power is required to generate interference.)
2. The "power under the curve" contained in a particular interference. (Actually, though, the *voltage* is most important here—these are radio frequencies with very limited current capacity.)
3. The impedance of the circuit being affected. Here, a *lower* impedance is safer, due to the limited current capacity in any particular pulse. Or, stated another way, the power in a given pulse is equal to the voltage multiplied by the current, which is a function of the impedance in the circuit.
4. The distance between the interference source and the circuit affected in the case of radiated RFI, because the power of a radiated signal decreases by the square of the distance.

These variables can each span several orders of magnitude, making this an extremely complex problem. In fact, RFI can exist all the way across the radio-frequency spectrum. We term this *white noise* because there is no particular pattern to it. Computers are most bothered by noise frequencies from about 100 kHz to around 10 GHz or so. This is pointed out in Figure 1.2, which shows the required power decreasing as frequency rises. (See also Fig. 1.10, which shows a graphic of white noise.)

If we were to plot frequency versus the level of RFI required to affect a given peripheral, it would look very much like the plot shown in Figure 1.2. As frequency increases, the signal level required to affect the computer becomes less and less. The accompanying graph in Figure 1.2 is generic in nature and does not represent any particular computer or peripheral, but it is meant to illustrate that higher frequency range is the most destructive to a computer when considering radiated RFI susceptibility.

Electromagnetic Interference

In this book, we will define the interference caused by the high currents in electromechanical devices as "electromagnetic interference" (EMI). But actually, the phrase "electronic forces

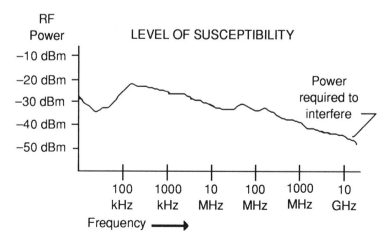

Figure 1.2 A typical microcomputer RFI susceptibility curve.

induced from (or as a result of) magnetic interference" would have been more descriptive, because EMI actually consists of varying magnetic fields from an electromagnet produced by current passing through coils. This magnetic field has a definite strength that can be measured. It can be mapped in "relative field strength," as shown in Figure 1.3. The relative strength of an electromagnetic field's effect in space falls off by the square of the distance as shown in Figure 1.3.

By way of review, you probably remember that a magnetic field expands around a wire or coil as current begins to flow through it. When current through the coil ceases, the magnetic field collapses. In the case of EMI from an electromagnetic device, this expanding and collapsing magnetic field will induce a corresponding voltage *into any conductor it "cuts" across*. The voltages induced by EMI are not coherent (contain no useful information) and cause intermittent computer problems when the conductor that is being cut by magnetic lines of force is part of the computer. This is the most common result of EMI. Other ways EMI can affect a computer are the *magnetic* effects on the magnetic recording devices used in computer systems, as pointed out in the following example.

Case in Point

I once had a field engineer tell me that he had an account that was "wiping out" one or two diskettes a week. He hadn't

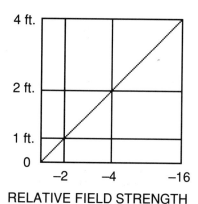

Figure 1.3 RFI field strength.

been able to figure out what was happening. Finally, in desperation, he came to me and the two of us made a visit to the site. But first, I made a visit to a sporting goods store and bought a hand direction-seeking compass. "What did you buy a compass for?" asked my friend. (Can *you* guess why?)

On arrival at the customer's site, several things became clear right away. The customer's diskettes were being kept in the top drawer of a metal slanted-top upright writing desk. Since this particular computer was small, as supermicros go, the customer had not done *any* site preparation. The computer had been recently installed in a room with no air-conditioning, open to contamination from the dirty repair shop beyond.

The first thing we did was take the compass and check around the "computer room," which contained another 30-in. × 60-in. metal desk, a chair, the metal upright writing desk, a heater, a telephone, and not much more. By holding the compass close to the nearby metal desk, I saw that it was magnetized. So was the metal upright writing desk in which the floppy diskettes were kept. While we were talking to the service manager of the dealership, I laid the compass on the desktop. As we talked, the lights suddenly dimmed and a great whirring, thumping sound began, actually vibrating the ceiling. "What was that?" I asked. "Oh, that's the air compressor upstairs," the service manager replied. I glanced at the compass, and it was *spinning wildly*. I pointed it out to my friend, and said, "There's what's wiping out all the diskettes!"

He could not believe what he'd seen.

It turned out the air compressor was powered by a 20-horsepower motor, which went on and off many times a day. When converted to AC line current, 20 horsepower represents a whopping number of amperes!

Using Ohm's Law, since

$$\text{Hp} = \frac{P}{746} \quad \text{and} \quad I = \frac{P}{E}$$

where P = power in watts
 E = voltage
 I = current in amperes
 Hp = horsepower

Then,

$$I = \frac{\text{Hp} \times 746}{E}$$

This means that a 20-horsepower motor running off 220 *volts* (V) would be drawing roughly 67.8 *amperes* (A) of current. If the motor were running off 440 V, the current still would be roughly 33.9 A. With that much current being switched on or off, it was no wonder a magnetic field was being generated that was strong enough to magnetize every piece of ferrous metal around. My friend was absolutely astonished. He had never seen a magnetic field so strong from an appliance before.

This was not, however, really unusual. It just so happened that the computer had been very recently installed into that room, with absolutely no planning or forethought about interference sources. No one had thought about the fact there might be an air compressor or other device nearby that could interfere with the computer. At my insistence, the computer was relocated to a clean, air-conditioned room—away from air compressors, heaters, fans, air conditioning equipment, etc., and the problem disappeared. If the diskettes had been properly protected in the first place, I probably never would have gotten involved. But eventually, this system would have suffered terribly and would have required excessive repairs and maintenance to keep it running. In fact, floppy-disk drives, power supplies, and probably the hard disk would have eventually required replacement—possibly with the complete loss of the customer's data.

Does this mean that preparation of the eventual installation site is the most important part of system problem prevention?

The answer to that question is "No." It's a stopgap, at best, when it is "after-the-fact" as in this example. You might say that "site prep" is the final act—necessary, but not the whole show. Reduction in noise susceptibility must be *designed* in, all the way from board-level design to integration and, eventually, site preparation and installation.

If proper and thorough computer power and ground design is done in the beginning, there should never be problems from a system catching a "zzaap" from an outside EMI source. But it must be done "from the ground up," all the way to the AC power plug and beyond. Once you understand all the modes and causes of computer "zzaaps," you can design them out and prevent them from plaguing your product for life.

Because induced voltages in a conductor (especially a ground) caused by a stray magnetic field are not "coherent," as we discussed earlier, they fall into the class of electrical signals we call "noise."

The Effects of Electrical Noise

About now, I think we should do a bit of theoretical troubleshooting, and analyze what really takes place when a computer stops responding to the keyboard. The computer is obviously no longer sensing key-strikes. But *why*? What *is* it doing?

Suppose we were to look at the central processing unit's (CPU's) main clock pulse with an oscilloscope. If the clock pulse is there, and it appears clean, then the next question would be whether the CPU is running (putting out changing address patterns). If so, then it is no doubt doing fetches from memory, the way it should. But if the CPU is running, and it seems to be executing a program, then *what is* the problem?

Let's postulate a hypothesis— what probably happened was this: At the particular instant when the "zzaap" came along, one or more *data bits* changed on either the processor's address or data lines. One single data bit would do it. This could have caused the CPU to vector to some illegal address, or even try to execute something other than a "real" instruction. At that point, the CPU could have either gone into a very tight loop, or totally off into Never-Never Land (as I call it), thereby "blowing up" the program.

With the very low voltage levels found in computers [especially CMOSs (complementary metal oxide semiconductors)], it is not at all hard for a stray electrical impulse to exceed the 2.4 V threshold of the logic for a "1." Or, for that matter, to "pull down" an input to the point where it falls *below* the threshold for a "0." Either of these has the capability of making an address or data bus "pick a bit" or "drop a bit." And one bad data bit for one read or write cycle is all it takes. Later in this book, we will examine the exact mechanism that causes this, and try to prove this.

Indications of a "Zzaap"

Bright "flashes" appearing across the screen, or the screen filling with "garbage" is a dead giveaway that the computer system is being "zzaaped" by power or ground noise. It just happens in this particular case, that those impulses affect the screen directly or are induced into the I/O (input/output) lines, and do not affect the central processor or memory address and data lines. If the system has separate video terminals, it probably means that the CRT (cathode-ray tube) terminal itself is being "zzaaped." The IEEE (Institute of Electrical and Electronics Engineers) has formulated a "standard" transient waveform. Figure 1.4 depicts an approximation of the noise transient that is allowed to appear on building wiring in accordance with IEEE STD 587-1980.

As can be seen, this is a fairly severe voltage transient, but can be expected from a normal building AC power line. If the system has not been designed to withstand this type of varia-

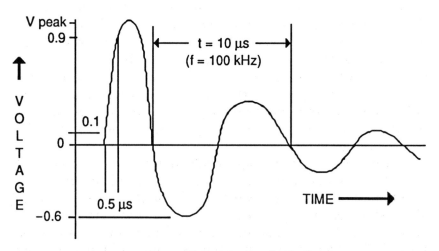

Figure 1.4 AC power "hits."

tion, it is guaranteed to show the effects—at exactly the wrong time.

When "error messages" from the system start appearing while the computer is "out to disk," it is a good indication that the disk or associated circuitry is being affected. This is the most destructive of all scenarios, because under certain conditions, it can alter data or even disrupt the format of the disk itself, causing complete loss of all data saved on it. In fact, this particular problem is one of the very good reasons behind doing a backup of a hard disk.

Designing out Noise

As every design engineer knows, the design of a computer system that is completely unaffected by outside sources of interference (noise) is not easy. Noise suppression (Fig. 1.5) must begin as a part of the most basic design phase, at component board level. From the layout of the board's power supply and ground to the placement of tracks, noise suppression considerations must be uppermost in the mind of the designer. For this is the "root" of the computer system. Only clean, reliable signals must enter and emanate from the computer board, or users will be plagued with continued reliability problems as a result. Problems stemming from a poor physical design and layout in the beginning can plague the manufacturer for the product life of a design.

As I stated before, and many computer manufacturers have learned the hard way, noise is a two-way street. If noise can get in, it can also get out. And with the increased FCC (Federal Communications Commission) requirements for low

Figure 1.5 No noise is good noise.

interference emissions, this is a very important consideration. Consequently, care taken to prevent outside interference will also help to prevent illegal emissions.

To be able to design out interference from ESD, EMI, and RFI, we must have a very basic understanding of each. The medium through which this interference normally travels as RF or magnetic waves is usually air. But eventually, RFI or EMI will be induced into wiring or conductors as an analog noise voltage. Airborne (radiated) interference mechanisms are completely characterized in Chapter 4. Chapter 2, on the other hand, takes the opposite approach, and covers conducted noise.

To give the reader an idea of how badly the above factors can affect system reliability, let's take a look at some "real-life" examples.

Common Errors

Because of choices of hardware available, very expensive vertical applications software written for a specific value added reseller (VAR) often no longer sells well as a "stand-alone." Lots of software houses in the vertical application software business have been forced to turn integrator to survive. Nearly every software-house-turned-integrator makes the same mistake: the assumption that any piece of digital hardware will work with any other, and will run in any environment. Actually, nothing could be further from the truth. The most common outcome of this kind of thinking is for the integrator to order the cheapest hardware available that meets his particular software operational requirements. Now, there is nothing wrong with trying to save a buck. But the old saying "you get what you pay for" was never truer. The problem is, the buying decision is very seldom made by an experienced engineer. The buyer may never consider whether the different parts of the system being put together are really compatible. Or, even if they are, whether the trade-off of price versus quality is worthwhile.

Here's a case in point. Several years ago, I worked for one of these software-houses-turned-integrator. The fact that a particular CRT was cheap and had the correct number of keys whose functions fit the requirements was uppermost in importance to the management. Circuit design and grounding were of no importance to them. They were so excited about having a software product that fulfilled an immediate need that they totally lost sight of hardware reliability—the lack of which caused unimaginable grief. In fact, it eventually caused the company to go bankrupt. They went broke trying to keep the systems they had already sold running.

Typical of the type of problems arising from this kind of thinking is one that arose from the internal grounding in a CRT terminal. An engineering order (EO) was issued to repair what *should never have been a problem to start with*. Furthermore, critical resources in the systems house itself were diverted to repair the problem, which escalated costs of operation. The following is a description of the problem and the fix.

Problem Description

This problem concerned the lack of a good common ground network inside the CRT terminals. It was caused by three things:

1. The manufacturer, for some unfathomable reason, decided to tie pin 1 (*chassis* ground) and pin 7 (*signal* ground) of the DB-25 RS232 connector *together* to *chassis* ground rather than connect pin 7 to signal ground alone and use pin 1 for the shield—pin 1 of an RS232C connector is supposed to be the green-wire AC chassis "protective" ground. But pin 7 is designated as the logic signal "return" and should never be tied to the green-wire chassis "protective" ground.
2. The chassis inside the CRT that served as a common ground was made of zinc-anodized steel. This zinc anodizing is not a very good conductor electrically. Therefore, it has been common practice in such cases to scrape or sand the zinc anodize off any sheet metal where a wire is to be connected. This had not been done on an unknown number of CRT terminals.
3. To prevent the grounding screw from loosening, the

manufacturer, in the manufacturing process of the CRT, deposited a locking compound on the screws. The grounding screw was one of these.

This resulted in the locking compound (itself a very good insulator) running into and between the wire-connecting terminals, the screw, and the chassis ground plane. This totally insulated each wire connection from the others, and both wires from the ground plane. An enlarged side-view drawing of these wires and the grounding screw appears in Figure 1.6. The original EO contained an overall view of the CRT terminal with the location of this screw pointed out. It also included a cross-section drawing of the joint that was causing the problem. Both of these drawings are included here to help clarify the nature of the problem.

This kind of quality-control problem can exist in any production line. It is not unique to the particular product used in this example. It happens every day, and sometimes it is due to an unlucky fluke during the part's manufacture, but at other times it is accidentally "designed-in." I am including a hypothetical EO for correcting the problem, to point out the extent of labor costs that just a small oversight on the part of a technician or engineer can incur.

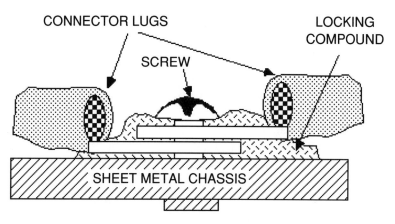

Figure 1.6 Grounding error.

The Repair

To repair this problem alone required the following steps on each and every CRT terminal (shown below as though you were reading page two of the EO):

Engineering Order # 1

Page 2

1. As depicted in Figure 1.7, loosen the four screws at the rear of the terminal case (two at the top and two at the bottom). These are "captive" screws, and will stay in the rear half of the case for reassembly later.
2. Pull the rear half of the case off carefully, and set it aside.
3. Using Figure 1.8 as a guide, locate the grounding screw in question.
4. Remove this grounding screw, being careful not to break any of the wires connected to it.

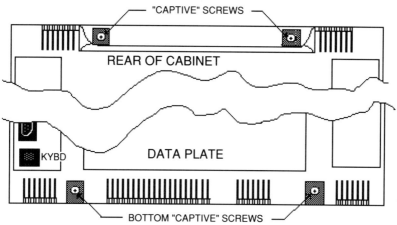

Figure 1.7 Rear cover mounting screws.

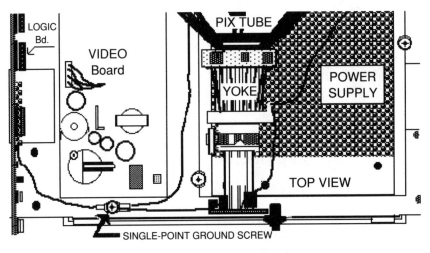

Figure 1.8 Location of the grounding screw.

5. Remove the wire spade connectors from the screw very carefully. Clean the top and bottom of these connectors at the point where they make contact with each other or the screw until they gleam.
6. Clean the screw, paying particular attention to the underside of the screwhead. File if necessary, to get a clean metal surface under the head.
7. Now scrape or sand the surface of the sheet-metal chassis at the point where the ground screw enters, until it gleams. Remove all metal filings and dirt carefully.
8. Replace the wires under the screw to the grounding screw hole and tighten. Check for continuity with a VOM (Volt-Ohmmeter) to assure a good connection to ground.
9. Replace the rear cover of the CRT cabinet; be very careful not to crack the case. Tighten the four "captive" screws.
10. The repair is now complete. Record or check off the serial number of this unit on the list of CRT serial numbers obtained at the beginning of this EO.

Time to repair per unit is estimated to be 0.5 man-hours.

As you can see, the time required to repair several hundred of those CRTs converted to a good deal of money. Yet the systems house preferred to absorb the cost, rather than forcing the original equipment manufacturer (OEM) to make the repairs or, at the very least, absorb the cost of them.

Being a systems house in a vertical market, they were more interested in serving the customer's immediate needs than in long-range planning. They hired field representatives who came from the automotive field (since that was their market) rather than computer people. The inevitable result was that field reps would order a replacement part to satisfy the customer without being absolutely sure what the problem really was. Shipping costs began to skyrocket. With only some 140 systems in the field, they were spending over $70,000 per month in shipping costs, to replace hardware that in many cases wasn't bad to start with. "Lock-ups" became a major problem. Customers were doing system reloads as often as every 30 minutes. Management was literally "giving away the farm" to stay in business.

All this had come about as the result of several factors:

1. Lack of environmental requirement specifications for installation of their systems, which caused many additional problems, such as
   a. Increased probability that a system would be "zzaaped" by environmental noise
   b. Decreased individual component reliability by overstressing parts beyond factory specifications, especially RS232 drivers and receivers (see item 5, below)
   c. Exposure of delicate computer components to all sorts of other environmental hazards such as smoke, dust, contaminates, fumes from solvents, etc.
2. Insufficient testing of component compatibility before making purchasing decisions
3. Incorporation of improperly designed peripherals simply because they were cheaper
4. Choice of an unreliable mode of transmission for I/O signals (See discussion of RS232 vs. RS422 in Chapter 3.)
5. Stretching the limits of RS232C, in terms of cable

lengths, by using cables as long as 350 ft (7 times the EIA specification)

As if all these were not enough, they further complicated the problem by trying to run a field-support effort with untrained personnel who knew nothing about computers and could not diagnose system problems.

Over 6 months of research was spent finding all the causes of system "lock-ups" and making changes to increase system reliability. These changes were met with lots of opposition from the field. After several months of pure research and my visiting customer sites, diagnosing problems and recommending repairs, they began to see the reliability increase and "got behind" the effort. After more months of doing site preparation, running tests, writing EOs and product evaluation reports, etc., the system hardware was completely reliable.

By that time, hardware problems had become so scarce that the software problems began to surface. In fact, the programmers were finally forced to acknowledge and fix software bugs that had been masked up to that point. The software department had always claimed that all the system problems were due to hardware failures, and would never admit the software might be even partially at fault.

However, the results of past mistakes finally overtook them, and after having lost over ten million of the venture capitalists' dollars, they were forced to file bankruptcy—but not before they had built up a debt so huge that suppliers also lost hundreds of thousands of dollars.

Do not dismiss this story as the result of management stupidity, or even ignorance—although that might have been a factor. The moral to this story should be fairly clear. Proper design of the system, more effective grounding, and more rigorous testing of hardware that made up the system could have changed this from a story of failure to one of success.

Other Factors

From an overall computer system viewpoint, location and routing of I/O cables can make a tremendous difference. As has been shown, the power from an EMI or RFI source goes down by the square of the distance from the source. Thus, moving a cable only 2 feet away reduces the interference level by a factor of 4.

By the same token, tying that same cable to a well-grounded structure member can reduce the interference even more. Electric motors can induce noise voltages into I/O cables, as can radio transmitters—even though the cable itself is shielded. This is especially true if both ends of that shield are connected to pins 1 and 7 and to chassis ground, as described earlier. Under those conditions, we suddenly have a very big ground loop.

The Ground-Loop Phenomenon

The "ground loop" is a peculiar situation that nearly always results in computer problems. As you know, a ground loop takes place when unwanted current flows through a ground from the origination point, through the shield ground (or some other path) to the remote peripheral, and then back to the point of origin. The mechanism works something like the following.

Let's say the peripheral chassis is connected to both logic ground and the shield of the RS232 cable (by connecting pin 1 to pin 7 in the RS232 connector). Ground currents will now travel from the peripheral via the shield (an electrically noisy path), to chassis ground in the computer, through the green-wire ground pin of the AC outlet, then through the noisy earth-ground path back to the peripheral. The green-wire-ground path is also "noisy" because of ground currents (which should not exist in the first place) and is subjected to any disturbances caused by other electrical equipment anywhere in the vicinity. When a ground loop exists, the computer is most likely to be bombarded by ground noise created by such noisy electrical equipment as drill motors and vacuum cleaners, which have a tendency to spark badly at the brushes when they are running.

If all this sounds confusing, take a good look at Figure 1.9, which may help to make it a little clearer. Also, we will get much deeper into this subject and take a very detailed look at "noisy" grounds in Chapter 4. Once you recognize the different

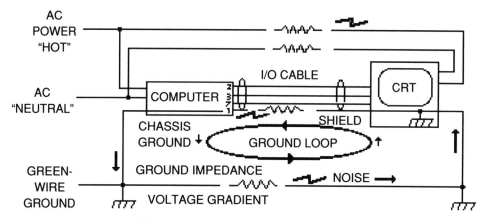

Figure 1.9 The ground-loop phenomenon.

noise effects and the symptoms associated with them, you will have your own crystal ball for insight into solving the various problems they cause.

For example, suppose you own a home computer that occasionally goes off the deep end into Never-Never Land. What can you do if it is not completely "zzaap-free" to the point where you can trust it for more than an hour at a time?—You should start by looking for evidence of ground noise. I once found "white noise" on the internal ground of a desktop "supermicro" computer that looked like that shown in Figure 1.10 . . . and *this particular system had an internal hard-disk drive.*

Ground Noise

We are going to describe a scenario that is typical although hypothetical. As you can see from the accompanying simulated oscilloscope presentation in Figure 1.10, the AC "white noise" on the ground of this computer runs the gamut of frequencies from 100 kHz all the way up into the gigahertz range. (Contrary to popular belief, it is actually the *higher frequencies* that freak out a computer.)

All computer equipment is subjected to some amount of this white noise. But they can be designed to tolerate it, up to a level

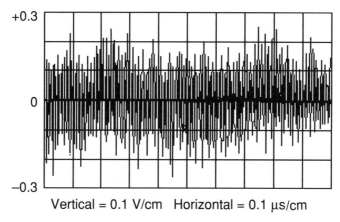

Figure 1.10 Ground noise.

of about 250 millivolts. This typical computer was bothered by noise because of several factors.

1. The manufacturer had built the case from expanded-core plastic with a sprayed-in conductive coating.
2. The conductive coating had been relied on to provide a low-resistance connection to several individual pieces of sheet metal, including:
   a. The card cage
   b. The power supply frame
   c. The floppy and hard-disk drives
3. The only ground wire was a single number 18-AWG (American wire gauge) wire "daisy-chained" to every piece of sheet metal instead of several wires radiating from a single-point ground to each piece of sheet metal.

To make matters worse, the resistance from one corner of the case to another, measured over 5 ohms (Ω). Too much ground resistance internally!

Using a commercial noise generator on an actual computer during several months of testing and investigative research, I completely duplicated the exact symptoms the users of this particular computer complained about, under controlled laboratory conditions. This allowed me to make controlled measurements, thereby isolating the noise components that truly affected the computer. After all that investigative research, I proved that ground noise at frequencies from 100 kHz up (especially those beyond 10 MHz) with more than 200 mV of amplitude will drive a computer completely "out of its mind."

Remember, I also proved that noise paths and RFI *are* a "two-way street," and any interference that can get *out* of the computer as emissions, can also get *in*. It stands to reason that if (1) RFI *is* a "two-way street," and (2) the computer is susceptible to RFI at the frequencies shown in Figure 1.2, then, the radiated and/or conducted emissions *from* that computer should exhibit a curve that is at least similar to that shown in Figure 1.2 for susceptibility. When plotted level-versus-frequency, any differences should appear as a *lower absolute level* only. And, as it turns out, they do!

A Light at the End of the Tunnel

But take heart—there *are* some things that can be done to make a home computer considerably less noise-susceptible. The truth of the matter is, a system designed throughout for low noise-susceptibility and coupled to a well-filtered power supply, then connected to properly designed peripherals through shielded twisted-wire RS422 cabling with the correct connections, is an unbeatable combination. In this way the user can be guaranteed that the system will not be susceptible to being zzaaped from the electrical noise that is all around us. The intimate details of why this is true will be covered later in this book.

Low Noise Susceptibility Design Requirements

A complete list of requirements that (when incorporated during design from board level to the system level) will pretty much guarantee low noise susceptibility, would be very long and complicated. But to impress on you the extent of the variables, here are some of the requirements.

1. PC (printed-circuit) boards should have a well-filtered power supply entry point, with bypass capacitors at

every IC, and protected with ferrite-bead feed-throughs at signal inputs and outputs
2. Separate power feeds to each IC row (not daisy-chained)
3. A bypass capacitor with short leads for every IC
4. Short ground-return paths with a large central ground plane
5. All sheet metal grounded to a single-point ground, from separate wires of at least 12 AWG or better (no daisy-chains)
6. Active outputs terminated by at least one input or a pull-up resistor, with interconnect tracks as short as possible
7. A mother board backplane that contains an integral ground plane
8. Design flat-ribbon cables so that every-other conductor is a ground wire (see example at the end of this list and in Fig. 1.12)
9. Hard disk and floppy disk drives well shielded, with the shield connected to ground
10. System processor enclosed in a well-grounded metal box
11. All I/O cable shields connected to metal or plated connector hoods, and grounded at computer end only
12. Power supply input well filtered, with entire supply shielded to prevent EMI or RFI being radiated

This list is not complete by any means, but it serves to indicate the number of things that must be taken into account in order to "design in" low noise susceptibility in a system.

As pointed out in item 8 of the list above, the diagram in Figure 1.11 shows how the "every-other-wire-a-ground" rule should be applied to flat-ribbon cable connections. In the PC-board view, the even-numbered row of pins are connected to a single ground track to ground every-other wire. The ribbon cable connector has the same alternating signal and ground arrangement. This provides maximum isolation and shielding between signals, and provides minimum cross-talk between signals.

A number of external criteria must also come into play when

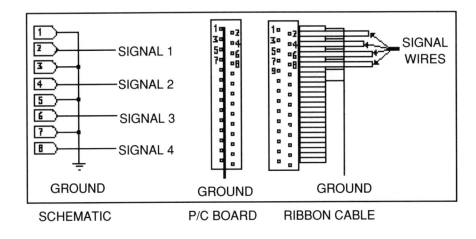

Figure 1.11 Ribbon cable grounding: the "every-other-wire-a-ground" rule.

considering the environmental aspects of installing a computer into a facility.

1. Peripherals that are designed so that logic ground and analog ground(s) are not mixed[2]
2. Environmentally sound site preparation, including:
    a. Separate, isolated-ground wiring from a clean, dedicated AC power line
    b. All peripherals plugged into dedicated, isolated-ground AC outlets
    c. No high-noise AC-powered appliances or equipment allowed near any computer equipment
    d. All I/O cabling shielded and routed completely away from any high-current power wiring, telephone cables, radio transmitters, or other potential noise source
3. Large volume of clean, air-conditioned air, held within 65–72°F, and humidity held within the range of 40–60%

---

2. The difference between the analog and digital ground is explained in Chapter 3.

Whether you are designing an individual processor or memory board, implementing a computer utilizing separate boards to build a system, integrating hardware and software, integrating a computer system into a facility, or whatever, just resolve to make "no zzaaps" a requirement for anything you design and/or implement. The whole world of users out there will appreciate it.

Satisfying all the listed requirements in the foregoing list will guarantee a trouble-free, reliable computer that will run for long periods of time with very little maintenance. And this is the goal of all computer manufacturers—right?

Probably, yes. But a hard look at all these requirements should convince you—it will not be cheap. Good quality is never cheap. However, care taken at the design stage will pay off handsomely in the long run. It may be tough to convince management of this, as we all well know. But they must be convinced, for your reputation is at stake!

There are probably more computer crashes caused by poor noise susceptibility in computers installed in a poor environment, than from any other cause (with the possible exception of poor programming). Design of the system must start from the power supply up with noise susceptibility and emissions in mind.

Remember, the FCC requires low emission from all computers and peripheral equipment. And designing for low *emissions* will make a large contribution toward low noise *susceptibility*.

You may have to sell management on the requirement for low noise susceptibility. If they are not sold on it, you may find your design does not weather well in an area of high thunderstorm activity. Nothing is harder to maintain than a system that is not designed to withstand the rigors of a poor environment. And believe me, the word gets around about unreliable equipment. More than one computer manufacturer has "bought the farm" and gone bankrupt from a poor reliability reputation.

CHAPTER 2

AC Power Variations

Now we shall take a guided tour through transients . . . AC power's fascinating "rogue's gallery" of surges, sags, glitches, spikes, noise, and power factor effects. (You may even come up with some yourself that we have not mentioned.) We will also look into the mechanisms that cause these, the effects of each one, and finally, how to design them out.

In every circuit two reactive elements, inductance and capacitance, manifest themselves whenever current amplitude is changing. According to Ohm's Law, if impedance (Z) remains constant, then a change in applied voltage must bring about a corresponding change in current. This changing current can wreak havoc in a computer. Not always, however, does the input impedance remain constant. This strange effect is covered later in this chapter, under the heading of "Power Factor."

Power Transients

In nearly every one of the following discussions on the different classes of voltage transient changes, we are speaking about the commercial 60-Hz AC line voltage. The reason for this, of course, is that nearly every small- or medium-sized computer is powered by commercial AC power with all its associated annoyances. They are therefore subjected to all the transients that commercial AC power can bring. The following is a description of the several different classes of changes in input voltage.

Power Surges

A power surge, in the case of alternating current, is a fairly sudden increase in the amplitude of the peak AC voltage, taking place over very few cycles. This rising voltage remains at its elevated value for only a short time before falling off again, usually to its normal value. What the load does with this change in voltage depends on the design of the load.

The sine wave shown in upper trace of Figure 2.1 is a sketch of a "normal" AC voltage sine wave. The peaks are all the same amplitude throughout, hence the peak voltage is not changing. However, in the lower trace of Figure 2.1, the amplitude of the peaks first increases, then falls off again. This constitutes a "power surge."

Sometimes a power surge can be seen—in the form of a sudden "flaring up" in the level of light given off from an incandescent lamp powered from the line, which then returns to its former level almost as quickly.

An AC power surge follows the same laws as any other electrical or electronic phenomena. That is, as long as the value

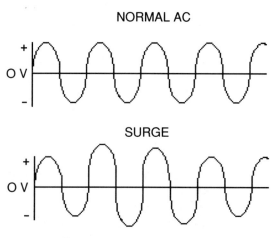

Figure 2.1 AC power "surge."

of peak current (I_p), and peak voltage (E_p) are constant, then the power under the curve is

$$P_p = E_p I_p$$

where P_p = peak power in watts
E_p = peak voltage
I_p = peak current in amperes

If, however, the voltage and current are *not* constant, the following applies:

$$\text{Average power, } P_{av} = \frac{1}{t} \int_0^t iv \, dt$$

where P_{av} = average power in watts
i = current in amperes
v = voltage in volts

Average power is different from apparent power. This will be explained further later in this chapter under "Power Factor."

Causes of Surges

Surges originate from several sources. At the top of the list of these are *switching transients*, caused by the "switching on" of appliances such as refrigerators, air compressors, etc.—anything that causes sudden increases in the current drawn from the line.

Another major source of surges is indirect lighting (fluorescent lights, mercury lamps, etc.). When these "flicker" as they turn on, they cause current variations and voltage transients, which can be reflected all the way back to the incoming AC power line. Street lamps with light-level sensing (which are usually of the mercury vapor type) start up individually at the first dusk. The electronic sensing/starting circuit causes the lamp to "arc over" internally in starting whenever the light level gets below that particular lamp's threshold setting. The starting arc also causes both conducted and radiated EMI and RFI, as well as a surge or sag on the line.

A surge is seldom a pure voltage surge, but carries with it

nearly every kind of disturbance we talk about in this book. This is because a voltage surge causes a corresponding current surge in the circuit (if the input impedance remains constant or decreases). A sudden increase in current causes a resulting change in the magnetic field surrounding the conductor, which can induce voltage changes in nearby conductors. Thus, any change in voltage is propagated as it brings forth a change in current, which causes a change in the magnetic field, which induces a voltage in other conductors, and so on.

The wiring leading to the foregoing surge sources becomes a very good radiating antenna. Both conducted and radiated EMI and RFI can be "broadcast" or spread from them over a large surrounding area. The more current these devices consume, the more power contained in the EMI and RFI from them. The EMI and RFI from these sources usually travels through the air, inducing noise currents into every conductor in their path.

Some circuits are more immune to this type of interference than others. The two main factors here are (1) the load impedance of the circuit involved and (2) the distance that circuit is from the source of the disturbance. Generally, the lower the line impedance, the less affected that circuit will be. And the power of radiated radio-frequency energy decreases by the square of the distance; so the further away the source of disturbance, the less it can affect a nearby circuit.

Some even less controllable sources of surges are the "natural" disturbances, especially *lightning*. The exact mechanism of any particular lightning-caused surge is usually unknown, due to the short duration and the unpredictability of that particular strike. However, it will probably be one of those on the following list.

1. Lightning strikes a primary circuit, injecting high currents into that circuit, which generate high voltages by flowing either through the load of that circuit, or by causing voltage gradients to exist in the ground path itself, due to a high ground impedance.
2. A lightning strike may totally miss the power line, and instead hit a nearby object. This sets up a varying magnetic field, which induces correspondingly varying

voltages on the conductors of nearby buildings or primary power lines.
3. A direct strike to ground nearby causes high currents to flow through the earth in every direction, setting up common-ground voltage gradients, thereby affecting anything connected to that ground.
4. A high-frequency voltage discharge takes place across lightning arrestors in the path of a strike. This disturbance is propagated both as conducted EMI via wiring, and as RFI and EMI radiated into the air itself. This discharge likely produces frequencies from the low kilohertz range into the gigahertz frequencies. At the currents involved, the power in these can be as high as megawatts.
5. Lightning may strike the secondary, or house-current-carrying circuits directly, and very high currents can be involved. Very high currents, indeed! If the path is of high enough resistance, the currents may actually vaporize whatever is acting as the conductor, due to the great amounts of heat generated. The mechanism here is the great concentration of current, heating the conductor due to its surge impedance.

A "Postmortem" of a Surge

To be able to relate the exact cause of one particular surge to its direct result requires information about the exact circumstances before and after that surge. Many times that information is not available immediately or at all. Therefore, insight into the *modus operandi* of a surge is needed. That's what we are about to do. Let's map out one particular voltage surge caused by lightning and its cause and effect from start to finish.

Lightning!

It was midsummer in a small town in Oklahoma. And, as everyone knows, there is no better place in the world to witness a lightning strike's force first-hand, than Oklahoma in the summertime. It was dark outside, and we were standing near a window, watching one of those famous "crash and boomer" evening thunderstorms. The very intensive downpour of rain, replete with wind and the ever-constant flash of lightning and crackling roar of thunder was very exciting.

Every once in a while, a brilliant, flickering flash would light up the night sky, followed seconds later by the reverberating boom of thunder. By counting the seconds from the flash of light to the clap of thunder, you could get an indication of how far away it was. And it was getting closer all the time.

Suddenly, a very bright bluish streak of light flashed from a nearby cloud to the roof of a house just down the street. The air was split with a tremendous thundering crack, causing such a rush of air that it could be felt a half a block away! A two-feet section of the roof of the house literally blew up near where the roof peaked. Even though the lightning had actually struck the roof of the house, all the lights went out for blocks around. As our eyes adjusted to the sudden darkness, we could see the roof of the house had caught fire. We ran to our car and drove down to the house, where the occupant was just running out onto the porch to see what all the uproar was about. He obviously didn't know his house was on fire! When we informed him his house was burning, he immediately called the fire department. They arrived very quickly and saved the house.

The point of all this is that the "lights went out, for blocks around." Why did a strike to the roof of a house cause other lighting circuits to lose their source of power? Later investigation proved that the strike had, indeed, used the roof as a point of entry. The high current through the wet shingles caused instant internal heating, turning the water to steam, and the shingles literally exploded. The lightning had then jumped to electrical wiring (which presented a much lower impedance to ground) once inside. The wiring, being vaporized by the tremendously high currents, had started the fire, along with the intensely concentrated heat caused by resistive heating of the wet wood.

Effects of Voltage Surges

A direct effect of the lightning strike described was of course, a huge surge on the house wiring circuit. That surge, traveling to the secondary winding of the nearest step-down transformer, was "stepped up" by the action of the transformer to the primary circuit *(until it blew out)*. This even higher voltage surge and transformer failure then plunged the local area into darkness.

Surges in the voltage of AC power circuits, either primary or secondary, can cause malfunctioning, or even failure of electronic equipment, particularly computers. Even though surge voltages have been measured and their effects observed, the exact waveshape and the resulting energy content of any particular surge are less well known. Several devices have been invented to prevent power surges from destroying computer equipment. But most of them have their own shortcomings, and some even cause secondary damage effects due to EMI, etc.

Power Sags

A power sag is defined here as a sudden decrease in line voltage—covering several cycles, with an eventual corresponding increase back to the nominal value. In this respect, it is an exact opposite of a surge. However, a sag can last for hours (in which case it is called a "brown-out"). At the end of a sag, however, there can be effects caused by the sudden return to "normal" voltage levels. The sudden increase back to normal voltage actually constitutes a surge, and therefore affects a computer in the same way that a surge does.

Figure 2.2 shows a voltage sag, and points out that *both* halves of a cycle are affected.

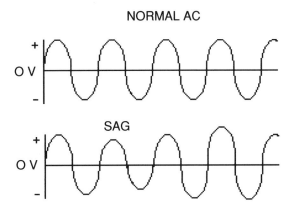

Figure 2.2 A voltage sag.

As a rule, a sag in voltage is not as destructive as other power problems—unless it is sustained over an extended length of time at low voltage levels (below 90–95 V in a 120 V-AC circuit). While it is true that lots of computer equipment will continue to function at this low voltage, the main effect of an extended minor voltage "sag" is a sustained increase in the amount of current required to maintain a given level of power. This increase in current mainly affects computers at the power supply itself, causing overheating of power supply components. Under certain conditions, it can even cause overstressing of a component to the point of failure.

Power Spikes

Our next "guest" is a very unruly ruffian known as a "spike." Of all power noise, the spike can be the meanest. Since a spike rides on top of a normal sine-wave voltage, it nearly always is conducted rather than radiated (although they can result from induction of radiated energy). An AC power spike usually exceeds the peak sine-wave voltage by a large amount; e.g., if the line voltage is 120-V-AC peak, a spike can exceed the line voltage by 150% or more. A spike is usually much less than a half-cycle time at the 60 Hz power frequency. In fact, it is too fast for some overvoltage-sensing devices to act to limit or clip its peak. Since its rise time is sometimes so fast, it can be radiated, causing RFI. A spike can be of such short duration that it can reach into the high megahertz range, if line lengths and impedances are just right. Due to the short rise-times of spikes, they propagate both as radiated and conducted EMI and RFI.

Figure 2.3 depicts a sketch of spikes compared to a normal AC power sine wave. It might be pointed out here that spikes can occur anywhere on the sine wave, including the zero-crossover point.

An extreme effect of a voltage spike coming in on the power line, if the amplitude is high enough, is insulation breakdown of the conductor carrying it. If this happens, there will be arc-over to the nearest path to ground, and a corresponding instantaneous short-circuit until the energy contained in the spike is dissipated. The arc-over can bring on secondary effects, including EMI and RFI, of both the conducted and radiated variety.

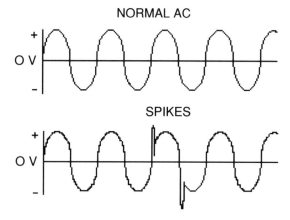

Figure 2.3 Voltage spikes versus normal AC.

This form of RFI and EMI generation is particularly destructive if born within the confines of the computer itself. If the portion of the computer where the spike originates is not properly shielded, the result can overwhelm the normal defenses against spikes. A spike can also cause extreme ground-gradient effects inside the computer itself—introducing noise of its own into the computer's circuitry.

Power Glitches

Under certain circumstances. a spike can become or cause a "glitch."[1] A power glitch can be defined as the the opposite of a spike, i.e., a very sharp *decrease* in the normal swing of the voltage. Glitches occur in very short time duration, much less than a half-cycle at AC line voltage frequencies of approximately 60 cycles. The accompanying sketch in Figure 2.4 depicts very simplified power glitches.

Bear in mind that there are other types of glitches beside power glitches. A "glitch" can also be defined as a very short-duration positive-going or negative-going change in voltage value on any conductor. This could also apply to computer signals on a data bus. In fact, except for a much shorter time

---

1. Although there are several uses of the word "glitch," for purposes of this book, we shall define it as a negative-going (decreasing voltage) spike.

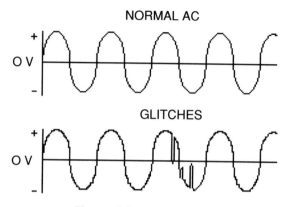

Figure 2.4 Power glitches.

duration, this is the exact mechanism by which a glitch causes a computer to "drop a bit" or "pick a bit" on a data or address line.

Suppose an address line (which is digital) happens to be high during an address cycle, and a negative-going analog voltage impulse or glitch of sufficient voltage is impressed on it. The result could very easily be that the address line is "pulled low" for a critical part of that time, and mistaken by the computer for a "low" or "0." The result: a "1" is mistaken for a "0," changing the state of a data bit. This can explain many of the elusive intermittent problems we see in computers. Since this kind of problem is unrepeatable, the rule of thumb might be: "if it's unrepeatable, it is probably a glitch on the power."

A glitch on the power line can also have a large corresponding spike that accompanies it, due to overshoot in the voltage during recovery. By the same token, undershoot following a spike can generate a glitch. In fact, spikes and glitches often take place as a mix of both.

Power Noise

All the foregoing effects could be lumped into one broad category: *noise*.

Sometimes radiated EMI or RFI, when encountering AC

power wiring that is not "loaded" too heavily by electrical devices, is converted by inductance (the "antenna effect") to a voltage of varying amplitude and frequency. These effects are usually exhibited as noise riding on top of the AC power sine wave, as observed on an oscilloscope and shown in Figure 2.5.

However, these noise voltages occur in two modes, *common-mode* noise and *transverse-mode* noise.

Common-Mode Noise

Common-mode noise is noise measured between *both* of the AC lines to ground, which can be caused by EMI or RFI being radiated from some other device (or phenomenon) and being picked up by the antenna action or induction of the AC power lines and other wiring. It can also be (and most commonly is) caused by electrical noise appearing on the ground line as a result of ground loops or faults somewhere in the building circuit. A very good test for this ground noise is simply to read the voltage between neutral and the "green-wire" ground at the outlet with a simple digital voltmeter set to read AC volts. This voltage should always measure less than 200 mV AC at the outlet, or trouble with computer equipment will be a guaranteed result. That is, unless the computer is "hardened" against this level of noise.

Common-mode noise, in and of itself, normally poses no real

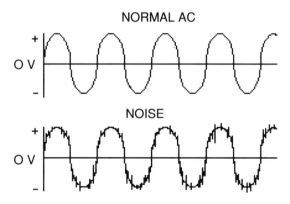

Figure 2.5 AC power line noise.

problem until it enters the internal power or ground of the computer itself. When it appears on the signal ground inside the computer, it can cause all kinds of weird data-related problems. This is because most computer logic inputs and outputs are referenced to ground.

When it crosses from one side of a power transformer to the other, common-mode noise becomes transverse-mode noise, which is *always* a problem.

Transverse-Mode Noise

How does it get there? The mechanism just described, whereby common-mode noise becomes transverse-mode in a power transformer, is one way. The transformer involved can be a step-down power line transformer on a power pole, or a distribution transformer in the facility. It could also be a transformer inside the computer itself, potentially the most dangerous and hard to control of all.

The transfer from common-mode noise to the less desirable transverse variety in a transformer results from the way the windings lay in the transformer in relation to the grounded parts of the core or frame. A voltage, due to interwinding capacitance, is induced into one side of the transformer output or secondary winding. This voltage, known as *transverse-mode noise*, will definitely affect computer operation. The solution is to shield the transformer to shunt the unwanted current off to ground before it can induce a voltage in the secondary. This is known as a *Faraday shield*. The amount of reduction of this current, or "rejection," is measured in decibels (dB).

An autotransformer, in which all windings are on the same side, has no common-mode rejection. The standard power transformer without shielding but with some isolation between windings will have some common-mode rejection. A Faraday shield increases this common-mode rejection to 50 dB. A computer-grade isolation transformer with boxed shields can increase this to 140 dB—a 100 million-to-one ratio!

Ground loops can be broken by inserting the right type of isolation transformer or line conditioner. But one must be careful when choosing a line conditioner to be sure that it is compatible with the problem at hand and really contributes to cleanup of noise instead of adding to it. This will be discussed in more detail later.

Power Drop-out

Power "drop-out" always presents a problem to computers, especially when the computer's power supply does not possess enough "ride-through" to "fill in the gaps," as it were. Ride-through is a general term relating to the amount of stored charge the power supply can call on to supply output voltage and current of the required level during an input power drop-out. This ride-through provides current to the supply in periods of low or no input voltage, usually for a very short time duration (possibly measured in microseconds). Power drop-out can extend from a glitch of less than a half-cycle to many cycles in duration. The longer the duration, of course, the more of a problem it can present to a computer.

In Figure 2.6, the power does not instantaneously drop from its sine value to zero, but rather "rings" up and down around zero. By the same token, when it comes back it is very noisy, and full of glitches and spikes. This type of drop-out causes the destructive RFI and EMI effects of more than just a sudden voltage drop and rise.

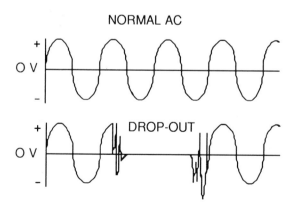

Figure 2.6 Power drop-out.

What happens to make the results so destructive? In the process of "going away," the power does not simply drop to zero. If it did, there would be no more adverse effects than simply losing whatever work had been done since the data was last saved, in the case of a small computer. Instead, the line voltage sine wave is punctuated by a very ragged high-frequency and high-amplitude surge–sag cycle that is as unpredictable as it is destructive. If a poorly filtered power supply is subjected to this type of abuse, it nearly always results in a hardware failure.

That is not the worst of it. If not disconnected from the line before power is restored, the computer may be subjected to another, even more severe high-frequency excursion of line voltage and current than that experienced during the "outage." This is even more destructive since the power supply must now draw huge "gulps" of current to try to reestablish a regulated output voltage. Voltages and currents under these conditions can easily exceed the specifications for the individual parts within the power supply, causing hardware failure. A device known as a "transient suppressor" or "surge suppressor" added to the power line may be of no value whatsoever in a situation like this.

The voltages clamped by the suppressor may be higher than the supply can handle. In other words, if the surge suppressor uses a class of voltage clipper known as *metal oxide varistors* (MOVs) of, for example, 250 V rating, the input voltage *can* exceed the specification for the power supply's input capacitors, rectifiers, or transistors by more than 50 V!

Or, conversely, the fact that the surge suppressor "holds back" the surge may cause it to "starve" the power supply for current, causing poor voltage regulation. One obvious result can be "punch-through"[2] of a capacitor, rectifier diode, or transistor. When this happens, the power supply may be doomed to destruction. Even if the supply is fused for slightly more current than it actually draws from the AC line, the fuse is much too slow to protect delicate electronic devices.

---

2. "Punch-through" is a term used to describe the breakdown of insulating layers internally in a semiconductor due to reverse voltage, causing a short circuit.

To sum up and review what has been covered up to now, Figure 2.7 shows each of the foregoing power problems as compared to each other.

The reader is referred here to an excellent article by William M. Kurple that appeared on pages 118 through 130 of the June 1985 issue of *Test & Measurement World* (an excellent source of information on this subject), entitled "Testing DC Power Supplies: Hidden Effects from the AC Power Source." Even though this article is several years old, the facts in it are still very much up-to-date. Rather than repeating the information contained there, I highly recommend reading this article.

For anyone interested in the use of surge-testing equipment, I recommend two booklets published as application notes by KeyTek Instrument Corp.: one presented at the 1983 IEEE International Symposium on Electromagnetic Compatibility and entitled "Application Note 111," and the other presented at the 1980 FAA/NASA Symposium on Lightning Technology, entitled "Applications Note 106." These two documents deal with spike/surge test waves and the generation and character of these waves.

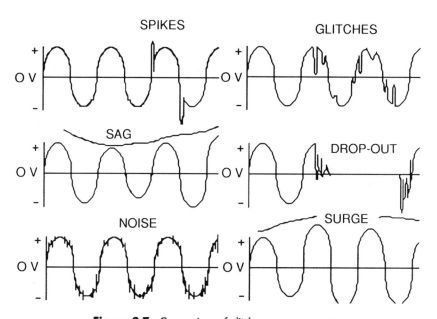

Figure 2.7 Comparison of glitches, sags, surges, etc.

Transient Suppression Devices

As computers shrink in size but grow more complex, the need for controlling or eliminating transients such as surges, spikes, glitches, noise, and RFI becomes even more important. The technology involved in controlling and eliminating these disturbances is a science in itself and has brought forth many types of contrivances that do the same job, but in different ways. Marketed as *surge suppressors* or *transient suppressors* by many different companies, they all use much the same components to do their intended job. The way in which the components are connected differs, but these transient suppressors embody the transient voltage surge arresting components we are speaking of here as the active parts or components that actually do the suppression of the transient. The rest of the transient suppressor is merely the circuit "glue" used to connect it to the power line. A complete list of all the different types of transient suppressors is beyond the scope of this chapter, but further (and far more detailed) information about them is given in Chapter 3.

Voltage surges and spikes on the inside of a computer require a special type of transient suppression device. That is, once the AC input power has entered the computer, it may still have surges or spikes present on it that can cause internal problems for the computer. As an "after-the-fact" cure-all for this kind of problem, some computer manufacturers resort to including surge-suppressing devices as a part of the computer's power supply. Others leave it to the user's discretion to buy one of the "off-the-shelf" contrivances.

But because you should be versed in the different types of transient suppressing components that are available and what each type is specialized to do, a short list of some transient- or surge-arresting devices currently in existence (and what each does best) is included here. The most commonly used transient suppressing devices currently available for use in surge suppressors include MOVs, zener diodes, gas-discharge tubes, and breakover diodes.

MOVs

An MOV consists of voltage-dependent, symmetrical resistors (known as varistors) which, according to the manufacturer perform in a manner similar to back-to-back zener diodes. When exposed to voltage transients, the varistor impedance changes from a very high standby resistance value to a very low conducting value.

This effectively clamps the transient voltage to a safe level. The energy of the incoming high voltage pulse is absorbed by the varistor, protecting voltage-sensitive circuit components. These properties in varistors have allowed them to be applied as line protectors where the capacitance and leakage currents are not significant.

MOVs can be connected *across* the AC line to prevent the input voltage from exceeding the voltage at which the MOV is rated. They can be connected across the coils of relays to absorb the sudden voltage surge created as the coil is deenergized and the magnetic field collapses.

The reader is urged to obtain a booklet available from General Electric entitled *Transient Voltage Suppression Manual,* available at any authorized GE dealer or OEM (original equipment manufacturer) supply house. This booklet totally covers transient cause, detection, and protection and includes a selection guide and specification sheets for GE-MOV™ varistors.

Zener Diodes

A zener diode operates on the well-known "reverse-voltage knee" principle. As reverse voltage increases, a point is reached where the zener diode begins to conduct current. When connected in series with a resistor to limit current, the voltage will remain at this "knee" even as current through it increases. This forms a very effective voltage regulating device. This device can be connected in many ways, and even used as a voltage clamp. Zeners are used extensively at low voltages to limit internal transients in transistor circuits. For the lower voltages and energy transients that are generated mostly from internal sources, zeners are very effective.

Further information is available from many sources, especially the device handbooks from the manufacturers themselves.

Gas Tubes

The application of the two components discussed above as transient protectors has been largely in the low voltage ranges that are not covered by gas-discharge devices.

The operation of a gas-discharge transient protector component is best understood by comparing it to a neon bulb, which relies on a slightly different effect, but yields a similar result. As soon as the applied voltage across the protective gas-discharge tube exceeds V_b (breakdown voltage), the current through the gas tube increases rapidly to values of several amperes or greater. The rate of rise in the current and the ultimate level reached is limited by the series impedance of the circuit. The voltage across the device at this time is a very low 20–30 V.

The reader is urged to acquire brochures on the above components from the manufacturers of these devices for further information. A very good source of information is a booklet entitled "Gas-Tubes®" by Lumex Opto/Components, Inc. of Palatine, Illinois. This booklet covers sources of damaging voltage transients and the several types of available transient voltage protection devices.

Breakover Diodes

A recent addition to this list of devices is the breakover diode. Touted to meet the need for reliable surge suppression to protect systems from lightning strikes and power supply irregularities by the manufacturer, the breakover diode is a bidirectional device that is claimed to shunt transients from either direction. It switches to a low voltage "on" state almost instantaneously at breakover and consequently can handle much higher impulse power, according to advertising claims.

These devices come in TO-220 packaging, and in versions for

120, 140, and 220V. More information is available from Amperex Electronic Corp., Smithfield, Rhode Island.

Problems Associated with These Devices

The fact is, a device such as an MOV, zener, or gas-discharge tube may cause more problems than they cure if not used properly. First off, they are all sources of noise . . . especially at voltages near the limiting point. However, used according to the manufacturer's application information, for the proper reason, they do have merit.

AC Line Conditioners

If proper environmental planning and system design is done in advance, there should normally be no need for AC line conditioning. But sometimes a user is completely at the mercy of a computer manufacturer and the power company, especially with some home or small business computers. For example, one may have no choice but to run a desktop unit from very bad AC power input. In this case, a desktop unit may find itself required to run with very bad power input. Such a situation is what an AC line conditioner was designed for.

Types of Line Conditioners

There are several types or classifications of line conditioners. They are classified according to the method used to control the voltage excursions they handle. For example:

1. Ferroresonant transformer devices (also known as "saturable core reactors"). These control voltage excursions by using the "saturable" feature of the special core material used for the transformer's core. An extra winding or two is added, wound in a way that allows it to "buck" or fight voltage transients. These are specialized devices, particularly suited for one type of transient control, which they do very well. These should be carefully tested under

actual operating conditions, however, since under some of these conditions these devices are totally unsuited and may cause system problems.

2. Switching regulators. These devices regulate voltage by switching in or out additional turns on the secondary of the transformer that it contains; i.e., if the voltage goes up, the regulator switches to a lower turn ratio by switching turns out of the secondary. If the voltage goes down it switches extra turns in until the voltage is back up within its tolerances. This type of line voltage regulation has several built-in problems, especially if used in conjunction with overvoltage suppression or sensing equipment.

3. Electronic switching regulators. This does basically the same thing as the switching regulator in item 2 above, except that it is controlled electronically. This method is a little smoother, but more complicated, more expensive, and more subject to failure. However, they do a slightly better job than the switching type and are worth it.

4. Electronic voltage regulators. Be careful not to confuse this device with 3, above. A true "electronic line voltage regulator" does just that. It *regulates* the incoming voltage electronically, not by simply switching turns in and out. The main drawback of this type of line regulation is that it is very expensive, and usually bulky (read also "heavy"). It also requires cooling or at the least good ventilation.

5. Uninterruptible power supply (UPS). A very short discussion of this device will be given here, since it will be gone into at great length later in this book. Basically, there are two types of UPS devices. First is the switch-over or off-line type, containing a battery kept charged from the AC line, but which does not ordinarily supply current to the load. At the time of a power outage, the UPS senses the lack of input voltage and switches over to the battery. Since it is not of the on-line type, this unit has several inbred problems, including the time lag of the switch-over itself. This delay can be seen by the computer as a power drop-out. The second, more reliable type is the

on-line UPS, which always supplies current to the load from a battery. During times when AC power is available, the battery is constantly being recharged. If the input voltage "goes away," the battery simply continues to deliver power, with no switch-over delay.

Last on this list, since it is not a true line conditioner in the sense that it dynamically changes the input voltage levels, is the *isolation transformer*. The characteristics of this device are such that instead of providing control over input voltage, it provides *isolation* from it. In other words, if you are bothered by a groundloop, which introduces noise into your computer, an isolation transformer is meant to break this ground loop. It may or may not do this, depending upon the exact situation.

One other device we have not mentioned so far, because it is specialized, is the lightning arrester. These are usually used in what is known as the "tornado belt" or the "thunderstorm area" of the United States.

The Lightning Arrester

A lightning arrester is usually cheap in comparison with the foregoing conditioners. It is made specifically for one purpose—arresting, or "shorting out" the high voltages and currents associated with lightning. One must be very careful when considering the use of a lightning arrester, however, for this very reason. The voltages and currents involved in a nearby or direct strike from lightning are nothing to play with.

If you are not thoroughly familiar with lightning and its odd behavior (although it follows all the "laws" of electricity that any other kind does, it is much harder to predict), do not attempt to install or even recommend one of these devices, for you might be "taking a tiger by the tail!" The currents involved in a direct lightning strike can vaporize ordinary wiring. Therefore, improperly installed, this device can represent a real hazard to both buildings and personnel. By the same token, if the device is not properly grounded through a sufficiently heavy conductor, it can build up dangerous voltages during a lightning storm that, if improperly "drained off," can either arc over and cause a fire, or if inadvertently touched, could kill a person. Also, the extremely heavy currents involved in a lightning strike can vaporize the ground conductor, causing a fire.

Lightning is covered thoroughly, with the exact mechanism given in Chapter 4 of this book.

Power Factor Effects

So far, we have not looked into the phenomenon that manifests itself as a difference between the *real*, calculated load current of a given load and the *apparent* load current read from current-monitoring devices. This difference between real power and apparent power is known as *power factor*, and is expressed as a decimal number related to the percentage of the difference, i.e.:

$$\text{Power factor} = \frac{P_r}{P_a}$$

where P_r = real power
P_a = apparent power

Any distortion of the AC voltage or current sine wave results in a less-than-unity power factor. (A power factor of 1.00 is perfect.)

Traditionally, a poor power factor was the result of a phase shift between voltage and current due to inductive or capacitive loads. These might include induction motors or capacitive-start AC motors, etc. However, in the case of switching power supplies, the poor power factor is usually the result of a load current waveshape that is not a sinusoidal wave yet is still "in phase" with the input. The problem is that in the case of input AC power to a switching power supply, the apparent circuit load impedance appears to change during an AC cycle, resulting in some very complex problems.

How could this be? What could possibly cause the input impedance to a power supply to appear to change? This effect comes about because most switching power supplies are designed with a full-wave input rectifier, with filter capacitors immediately following the input rectifiers, as shown in Figure 2.8.

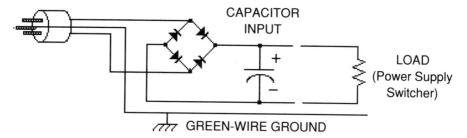

Figure 2.8 Capacitive-input switching power supply.

The result of capacitive input is that these switching power supplies draw current in great "gulps" that correspond to the input sine-wave voltage peaks.

This phenomenon is shown in detail in Figure 2.9, where the capacitor charge voltage level appears as the top trace, compared to the AC power voltage sine wave in the middle trace. The bottom trace shows the current requirements of the capacitive-input circuit.

Let's go through a few cycles of input alternating current on

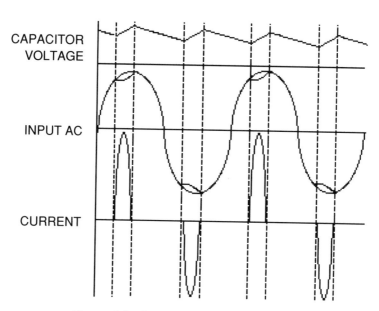

Figure 2.9 Capacitive-input current waveshape.

a typical capacitive-input switching power supply, and observe this phenomenon:

As shown in Figure 2.9, as the input AC voltage builds up and is rectified, it charges the filter capacitor to its maximum voltage (peak applied voltage, minus rectifier voltage drop). Then, as the input AC voltage again falls off toward zero, the capacitor, in supplying current to the switcher, becomes lower and lower in charge level (read "voltage"). As the AC sine wave reverses polarity and again begins to increase in opposite-sign voltage, the capacitor has discharged to a lower-than-peak voltage. The capacitor then begins to draw current once again to replenish its level of charge. This current is delivered in a large "gulp" in a very short, near-peak-voltage part of the input AC sine wave. Once recharged to maximum, the capacitor stops drawing current. As current is consumed in the power supply switcher, the charge level again drops off, and the cycle repeats.

This results in current peaks that are several times higher than a normal constant-impedance resistive load would draw. But they are of short duration. Hence the *apparent* power is much higher than the *real* input current, even though they are "in phase." Since power factor is the ratio of apparent power to real power, this effect results in low power factor. ("Apparent" power is defined here as apparent current times instantaneous voltage, which does not take "duty cycle" into consideration.) Real power is defined as

$$Ap = I_{ir}E_p,$$

where I_{ir} = Real input current
E_p = Peak instantaneous voltage at that moment in time
Ap = Apparent power

One of the design techniques used by some manufacturers to offset this problem is the passive approach, with an inductor in the input to the switching power supply, as shown in Figure 2.10. If the inductor, or input *choke* is large enough, the power factor can approach 0.90. The drawback to this technique, though, is that an inductor large enough to do the job is both bulky and heavy. A choke of sufficient size to do the job will increase the bulk of the power supply by 1.5–2 times the origi-

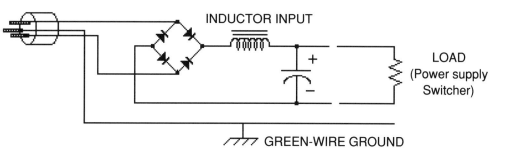

Figure 2.10 Inductive choke input.

nal size, and can triple the weight! This weight and bulk increase would be unacceptable in most cases.

There are at least two other alternatives: the filter input and the active power factor correction. These will both be covered in the following chapter. (This may tend to sound as though this subject is not worthy of further discussion. Actually, nothing could be further from the truth. I am simply deferring this subject to the next chapter, where I can give it the attention it deserves.)

CHAPTER 3

Surge Suppressors and Noise Filters

Many different types of surge suppressors, line filters, and other add-on equipment are available to owners of home computers and equipment utilizing embedded processors (dedicated-processor-driven equipment)[1], supermicros, and minicomputers. These can be purchased off the shelf at computer stores, electronics stores, and even at do-it-yourself supply stores. A typical small surge suppressor is shown in figure 3.1.

But if the truth were known, a good portion of these are simply not effective for most of the power problems found in the average small computer or embedded processor's environment.

Moreover, these surge suppressors could not begin to offer protection for the larger computer systems, or high-current-demand computer-controlled equipment[2] simply because the current demands from these are so great as to rule them out. Besides, the truth of the matter is, voltage surges are by no means the only (or even the worst) power problems a computer system can be exposed to.

---

1. Equipment such as microwave ovens, computer-type washing machines, computer-controlled stoves, irons, and toasters.
2. Such as computer-controlled electric irons, toasters, and combination convection and microwave ovens.

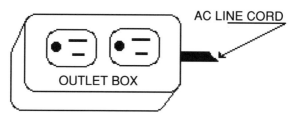

Figure 3.1 Typical surge suppressor.

Equipment Design and Selection

A large percentage of the so-called surge suppressors or transient suppressors found in your local computer or electronics store are simply over-voltage protectors utilizing metal oxide varistors (MOVs) and/or gas-filled gas-discharge tubes (discussed in Chapter 2) plus possibly a few capacitors, as shown in Figure 3.2. These surge suppressors should be viewed with a jaundiced eye and very carefully fitted to the job by observing their specifications versus the requirements.

"What's wrong with using MOVs or gas-discharge tubes?" you ask. Actually nothing, except that they alone may not completely cover all the AC input power problems that the average embedded-processor system, small business or home computer is likely to be exposed to. Remember from Chapter 2 that an MOV is a device that has a particular maximum voltage beyond which it will "clip," "short," or absorb a surge (depending upon how it is connected). The problem lies in the fact that the MOV only clips a voltage spike that exceeds that particular MOV's

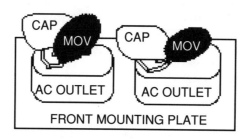

Figure 3.2 Surge-suppressor parts.

maximum voltage as long as that spike is *on top of the sine wave* (as shown in Fig. 3.3A).

As can be seen in Figure 3.3, the MOV will clip a spike that is of the same polarity (or riding on top of the sine wave), but probably will not "see" a spike that is of opposite polarity to the sine wave it is riding on (as in Fig. 3.3B). Many surge-suppressor manufacturers connect an MOV directly across the AC "hot" line to neutral, or from AC line to ground. In this method, the MOV simply shorts the line and absorbs the spike. This could result in a surge of current across the AC line. Under certain circumstances, this current surge could cause other, more damaging problems.

Since an MOV consists of voltage-dependent, symmetrical resistors that act similarly to two zener diodes connected back-to-back, any transient or constant voltage exceeding the particular voltage maximum for that MOV will cause the MOV to conduct. This effect can be utilized in many ways to prevent spikes from getting through to sensitive equipment. The trouble is, some transient/surge-suppressor manufacturers do not utilize the MOV in its most effective way. The effectivity of MOVs depends largely on the way they are used and connected into a circuit.

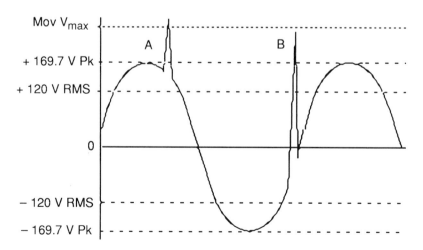

Figure 3.3 MOV voltage clipping action.

A Better Way?

Some of the more expensive and complex surge suppressors, spike suppressors, or transient suppressors may include a bifilar-wound toroidal transformer-type filter as well as perhaps additional capacitors and MOVs for each outlet, as shown in Figure 3.4.

This may be done in an attempt to provide inexpensive additional noise protection. These surge suppressors are considerably better, but for the most part may still be next to useless on the average home computer system installation, for two reasons:

1. They may not cover the "threat band" of frequencies most transients (or noise) occur in, because most destructive or disruptive noise occurs at some multiple of the computer clock frequency, above 1 MHz.
2. They may not "clamp" or "clip" at a low enough voltage to protect computer equipment because, based on a line voltage of 120 V RMS (root mean square) AC:

$$\text{Peak voltage} = \text{VAC}_{RMS} \times 1.414$$

where VAC_{RMS} volts = 120.

Therefore, at an AC voltage of 120 V RMS, the peak voltage would be approximately 169.7 V AC. This means the "clip" or "clamp" voltage for the MOV selected must be *more than* 170 V (120 × 1.414). But the nearest available MOV maximum voltage might be 250 V. If so, the AC voltage must rise to at least

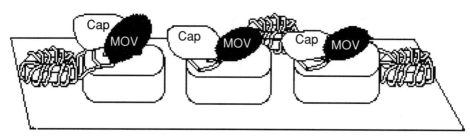

Figure 3.4 Additional toroid coils may be added.

250 V before the MOV will "clamp" or conduct! By this time, the computer may have gone up in smoke!

But why is there such a high demand for surge or transient suppressors, if they may not work for the real AC power problems a small computer might encounter? Several reasons exist for this. To begin with, the normal small home or business computer manufacturers do not (or cannot) provide the "ride-through" or filtering of input AC power that the computer really needs, for reasons of manufacturing economics.

Many computer manufacturers prefer to let the "third party add-on" manufacturers provide the extra filtering, conditioning, and even power backup. But if the truth were known, small home or business computers need the power input filtering and leveling devices the worst! Not only that, but the home computer users for the most part don't realize all the problems they can encounter from bad input AC power. So they don't realize they are being "short-sheeted," so to speak, until long after they have purchased the equipment.

The first indication they may have that something is amiss is the complete loss of data when a file (or files) is (are) corrupted. Or the computer may go off into Never-Never Land and must be totally reset. When this happens, they usually ask around until they find a more knowledgeable user or computer person, who may advise them to get a surge suppressor. This may or may not be good advice. The point is, anything done at this point in time is "after-the-fact" instead of "designed-in."

The Alternatives

So what are the alternatives? We are about to discover that there are quite a few. First of all, what if it were cheap enough at the point of design and manufacture of the equipment to "build in" the ride-through, spike prevention and noise filtering required to meet the demands of poor site preparation and allow the computer to keep right on computing?

As we have discovered several times before, "hardening" of a computer against EMI and RFI should be attacked at the lowest design level—board design and layout, and design of the physical power and grounding throughout the machine. This is the point at which it is least expensive to the user. Besides, if the "hardening" is designed and built in, there is no decision required on the part of the user about acquiring add-on equipment later that may or may not do the job. Manufacturers will ultimately get blamed for any failures or malfunctions their equipment experiences, whether or not the cause was external to the product. And if a "zzaap" happens often enough, the user may get disgusted enough to "bad-mouth" the product, or even pursue legal recourse.

Actually, many users take the intermittent "zzaap" as a matter of course, and grumble, gripe, and even eventually *reset* the machine and rebuild whatever they must. But I assure you—they convey their displeasure to colleagues and acquaintances who will remember it when it comes time to consider a computer product *they* may be in the market for. This form of advertising, we can do without.

Sometimes users will be driven to the point where they invest (wisely or unwisely) in a more expensive device to solve their problem. They have other products such as isolation transformers and voltage regulators that they can choose from if they are willing to pay the price. While not as low priced as surge suppressors, line conditioners are relatively inexpensive and are reliable. Voltage regulators, on the other hand, tend to be expensive, bulky, and heavy. Some line voltage regulators have line noise isolation and transient suppression designed into them, and some don't. Again, if the system had been designed with EMI and RFI in mind, the customer would not be at odds with the computer's environment.

Design Techniques

As explained in Chapter 2, a capacitive-input switching power supply introduces many problems including but not limited to RFI and EMI emissions, power-factor problems, and more, due to capacitive input (as shown in Fig. 3.5).

The cause of the problem is the way current is drawn from the AC line, also pointed out in Chapter 2. Figure 3.6 shows a comparison of a pure resistive load's current waveshape versus a switching power supply's current requirements. In Figure 3.6,

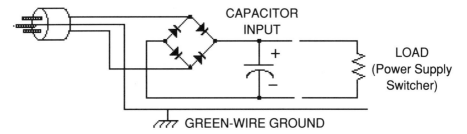

Figure 3.5 Capacitive input.

both the resistive load and the switching power supply draw 2.0 Amps RMS each. But as can be seen, the switching power supply current is drawn at near-peak voltage points on the input sine wave and the switcher's current comes in quick, high-peaked "gulps" that makes power factor appear somewhere around 0.65. This is true because the capacitor acts as a "voltage reservoir" and recharges to the peak line voltage each time the line voltage exceeds the capacitor's charge voltage remaining. When recharged again to peak line voltage, it stops drawing current until the two voltages are equal once more.

The *resisitive* load's current will follow the input voltage sine wave as shown in Figure 3.6, as long as there are no inductive or capacitive components. Thus, its power factor would be very close to 1.00.

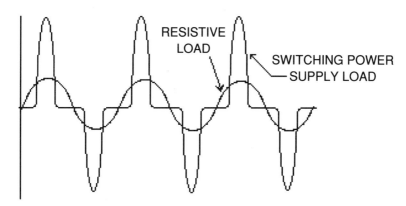

Figure 3.6 Resistive load current versus a switcher.

Let's examine the switching power supply input requirements versus the mechanism required to accomplish each of the requirements. With this information, you will be prepared to design and implement an input circuit for your exact requirements.

First, we are faced with the following set of requirements:

1. Weight: No increase in weight of the power supply beyond 10% of the original.
2. Volume: Volume of the power supply shall not be increased by more than 15%.
3. Cost: The increase in cost must be held to no more than 120% of the cost of the original.
4. Modifications required: The device must be a slight addition to an existing switching power supply to meet the desired specifications.

Power Input Device Specifications

From this set of requirements, we draw up a set of power input device specifications:

1. The device shall reduce the level of incoherent noise emission to less than 150 mV of noise measured between any two wires on the input AC line, at any frequency between 100 Hz and 100 MHz. (*Note:* This may require very sophisticated measuring equipment to detect.)
2. Noise amplitude other than ripple voltage on the output of any DC supply shall not exceed ± 15 mV peak-to-peak.
3. DC voltage delivered to input of the switching power supply section must not contain more than 1% ripple.
4. Not more than 10% increase in overall weight
5. Not more than 15% increase in volume
6. Not more than 120% of original cost

Does this sound like a difficult specification to meet? It would be, with the normal capacitive-input switcher. What's more, it would be with the inductive-input approach by itself. And it might even be with an active electronic power-factor correction.

Equipment Design and Selection / 67

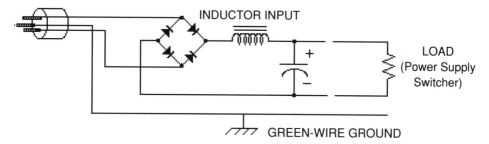

Figure 3.7 Inductive-input power supply filter.

Inductor Input

As mentioned in Chapter 2, an inductor at the input to a switching power supply as shown in Figure 3.7 can help considerably, but that help comes, unfortunately, at the cost of additional space and weight. An additional side effect is the slightly lower input voltage available to the power supply.

Filter Input

In the filter input technique, a noise filter is used in conjunction with the inductive input, at very little sacrifice in bulk or weight. The drawbacks to this technique are few and fairly painless. The biggest drawback is probably an increase in cost. (But you never get something for nothing, anyway!)

By adding a toroidal filter to the circuit shown in Figure 3.7, we could give the advantages of a choke coil inductive input—*without all the weight and bulk.* In Figure 3.8, the input trans-

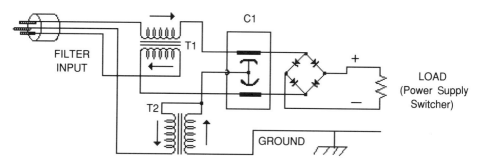

Figure 3.8 Toroidal transformer filter input.

formers shown have "lossy" toroidal cores, small in size and weight, but wound with two windings connected in opposite directions so that the fields oppose each other.

The circuit shown in Figure 3.8 requires only three additional parts (depicted in Fig. 3.9).

- A carefully wound pair of coils on a lossy toroidal core, through which the AC line "hot" and "neutral" wires pass in opposing directions
- An oppositely wound coil on a very lossy powdered-iron toroidal core, through which the ground passes from two directions (causing opposing fields in each pass)
- A integrated capacitor/inductor AC power filter, commonly called an EMI filter, connected to the hot/neutral transformer coil windings on the toroidal core.

However, the circuit in Figure 3.8 still has some problems. For instance, notice that the AC line end of the circuit has an inductor directly connected to each incoming wire. This tends to lower the overall voltage available to the switching power supply to a certain extent. Also, there is absolutely no voltage regulation or ride-through built into this circuit and available to the switching power supply.

What we need is a power source that will supply an even, filtered DC voltage to the switcher portion of the power supply and that will be "stiff" enough to provide constant voltage with varying current requirements from the switcher.

Enter the well-filtered, reserve-capacity, isolated-input

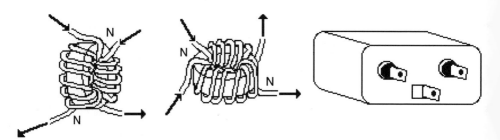

Figure 3.9 The three additional parts required.

power source. Simply rewiring the circuit in Figure 3.8 plus the addition of one extra component will provide a "stiff" well-filtered current supply with a constant voltage to the switching power supply. This circuit takes advantage of several things not available in one package with other schemes. For instance, it combines a two-way passive noise filter, which cleans up power noise that tries to propagate in either direction. Furthermore, there is provision for removal of either common-mode or transverse-mode noise by complete isolation of any noise that might appear on the ground. No EMI or RFI can escape to the AC line or ground from this circuit.

How It Is Connected

In this circuit (shown in Fig. 3.10), the "hot" AC wire goes first to the "load" side of the EMI filter (C1). From there through C1 to the "line" side, thence to the coil of T1, and through T1 in the direction of the arrow in Figure 3.10 to the full-wave rectifier.

The low (return) end of the rectifier bridge goes to the neutral "load" connection on the EMI filter C1. The "line" end of the EMI filter on the neutral side is connected to the opposite end of toroid coil T1. The opposing end of that winding is connected to the neutral AC line.

The green-wire ground from the input AC line is connected to the double-wound ground coil, T2. The opposite end of that winding goes to the ground post on the EMI filter, which is also

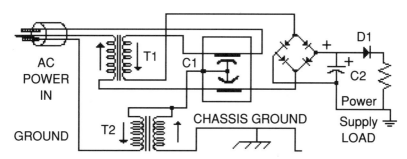

Figure 3.10 The ultimate input circuit.

connected to the same end of the other coil on T2. The other end of the second coil goes to actual chassis ground of the switcher. The EMI filter C1's case is insulated from the power supply's chassis ground. Remember both these coils and how they are connected, as it is not only important, but contains the very essence of why this circuit works so well. In both cases, use the "left-hand rule"[3] to be sure current through the windings of both toroids causes the windings to oppose each other. Be sure the instantaneous current through the two coils cause the "north" of each coil to be at opposing ends. This causes most of the spike or surge to be dissipated in the core.

How It Works

An explanation of the mechanism in this circuit is now in order. So that you can easily follow the discussion of the mechanism involved, I have included another view of the circuit as Figure 3.11 on the page where the explanation appears.

To explain this mechanism, let's take an instant of time when the input "hot" wire is just passing the negative peak and ready to become less negative in respect to the neutral wire as it drops back toward zero. (The assumption here is electron-flow from negative to positive.)

Let's assume that at just this instant, a negative-going spike riding on the negative peak of the sine wave comes along. The first component it would encounter is the hot "load" side of the EMI filter C1. The circuit inside C1 is complex, but consists of a circuit similar to that shown in Figure 3.11. This circuit could be characterized as two capacitive-input Pi-section network filters connected back-to-back. Due to the common-mode and differential-mode filters inside C1, the spike will be reduced by 6–50 dB, depending on the frequency of the spike in question.

If any energy remains in the spike after leaving C1, it next encounters the lossy toroidal transformer, T1. The rising leading edge of the negative-going spike meets inductive opposition in the winding of T1, which consists of only six turns of large,

---

[3]. The "left-hand rule" says if you curve the fingers in the direction the current is traveling around the core, the left thumb will point toward the magnetic "north" pole.

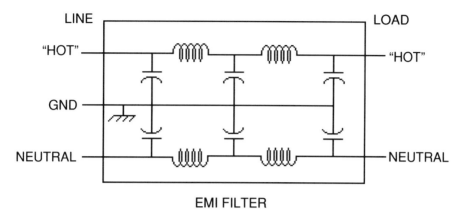

Figure 3.11 EMI filter equivalent circuit.

insulated wire (making it a very high-frequency winding). Energy, converted to a magnetic-field flux change, is transferred to the other winding, which is wound in the opposite direction on the other side of the core.

In the process of the transfer of this energy, some is consumed as heat in the lossy core. The rest is transferred to the other winding as an induced voltage that *opposes* the original impulse. Remember this—it is very important to the operation of this circuit.

From T1, the AC continues to the bridge rectifier and is converted to DC to charge capacitor C2. In the process of charging C2, the spike's energy simply adds to that already being stored in the capacitor. The accumulated charge on C2 is metered through diode D1 to the load (the switching power supply; see Fig. 3.12). Electron- flow returning from the load through the bridge rectifier next encounters the EMI filter C1, and the other winding of T1. Here any energy remaining in the spike is opposed by the voltage induced from the other winding, and loses additional energy through the lossy core. Thus the insertion losses throughout this filter for frequencies from 100 Hz to 10 GHz go to well above 140 dB! The insertion-loss profile of the entire circuit approaches the "notch" filter shown later in this chapter. It is near zero Ohms impedance below 100 Hz but climbs steadily, becoming higher as frequency increases. The

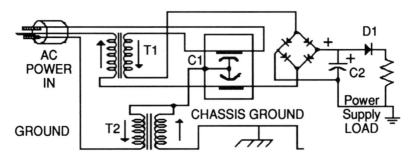

Figure 3.12 Final switching power supply input filter

characteristic impedance of this filter should be matched to the input impedance of the particular switching power supply's requirements. This can be done in many ways and proved by insertion loss measurements across the band of frequencies of interest.

Notice that in both the capacitive input circuit and the inductive input circuit, the green-wire ground was *not* a part of the power supply circuit, which is as it should be. However, in the filter-input circuits (especially the last one) the green-wire or chassis ground is an active part of the filter-input circuit.

This fact is one of the things that makes it so effective—we have isolated the AC power green-wire ground from the power supply's protective ground. Since this is all internal to the power supply, itself, there is no danger to personnel external to the power supply itself.

Another point not immediately obvious is that six turns of 12-AWG wire (for a filter rated for 3 A or less) wound on the toroid core, will exhibit a low impedance up to a few hundred kilohertz, which is exactly what we want. It will pass low frequencies and attenuate the higher ones (a low-pass filter). The cut-off point of the filter can be adjusted to as low as possible (preferably just above 100 Hz).

In order to show the virtues of this approach and prove that this design works, we will explore the methodology behind the exhaustive research and testing that was done on this device. Several hundred hours of research were expended to discover the exact types and frequencies of noise that bother the smaller computer systems the worst. During this research, it was also discovered that there are some hard-and-fast rules that must be observed in filtering out interference and preventing outside interference from getting in.

Those of you who are well versed in research testing may be bored to tears with the next few pages. But please bear with me, as I pass along as much of this accumulated knowledge as possible, and hopefully the insight that this knowledge brings with it.

Purpose of Research

In order to prove or disprove the value of a design it is always necessary to make exhaustive tests. These tests are always very time-consuming because they must be done in an extremely methodical and meticulous (read "boring") manner, in order to cover all the possibilities and variables. The following information is very important and will be needed to understand what comes later.

Research Methodology

The data accumulated in a research project is only as good as the method of testing that produced it. Its accuracy also depends a great deal on the test equipment used, and the way in which it is connected. We needed to collect a large amount of data in several categories in order to gather background information for the filter requirements of computers.

For our tests, it was of the utmost importance to prevent "background" noise and interference from contaminating the test results. To do this required a "screen room" or "quiet room." The equipment under test was installed inside the screen room, and all AC power entering the room was filtered by commercial EMI filters and line conditioners. Several good tests were devised to guarantee completely that all test results yielded usable, repeatable data, uncontaminated by background noise or conducted interference from outside the screen room. To do this required the prevention of outside interference from getting in. This was managed by filtering the incoming AC power and regulating it.

No cabling that formed a part of the equipment under test was allowed to exit the room. A "Zapper™" noise source was used to produce a very repeatable variable level source of RFI or EMI as required by the particular test being performed. The Zapper was equipped with several terminating devices that allowed the noise source to provide EMI or RFI for susceptibility testing of either the conducted or radiated variety. RFI and EMI emissions were measured in both the radiated and conducted modes from all equipment that was tested, with the setup described later.

Test Categories

Several different styles of tests were conducted, including the following:

1. RFI *susceptibility* tests of two varieties:
   a. Radiated
   b. Conducted
2. RFI *emissions* tests of two varieties
   a. Radiated
   b. Conducted
3. EMI *susceptibility* tests of both varieties
   a. Radiated
   b. Conducted
4. EMI *emissions* tests of both varieties
   a. Radiated
   b. Conducted

From this list, you can see that two forms of interference were the main target and thrust of the tests: RFI and EMI. Of these, two modes of propagation were examined: radiated mode and conducted mode.

Of these several types of interference, we looked for two completely different effects in the equipment itself: radiation from the device under test itself (emissions), and susceptibility to radiation being "thrown" at it from outside the device under test. Since noise is a two-way street, we were particularly interested in frequencies that showed up on *both* the emissions and susceptibility tests.

RFI Tests

To be assured that background RFI and EMI were not measured as part of the emission or susceptibility tests, "quiescent" background RF and EMI levels were tested first, with no AC power connected to equipment in the screen room at all. To do this, we rented a spectrum analyzer and examined the ambient RF levels in the room at every frequency to be examined. These were taken into account in test results. In addition, RFI and EMI were measured at the AC power input, at all frequencies of interest. These were eliminated through line filtering, shielding, or grounding techniques.

Some of the testing to be described in the next section may seem boring and redundant, but that is the nature of data gathering or research testing. Once the data has been collected, it must be analyzed and a meaningful information reduction performed on it. This sometimes means making the same test at several different levels.

Radiated RFI Susceptibility Tests

RFI susceptibility tests were run using both radiated and conducted RFI on many makes and types of computer and peripheral equipment. The methodology for both modes is explained in detail.

For tests involving radiated RFI susceptibility, the equipment under test was set up and connected to power. The RFI radiating antenna was set up at a measured distance, usually 39 in. or 1 meter (m) from the device under test. Extra care was taken to assure that the device being tested received RFI from one direction only, with any reflecting surfaces more than 2 m away. In this manner, the results could be easily tabulated and related directly to the angle and distance of the radiation source from the device under test. (Recall that power decreases by the square of the distance.) This setup would be recorded in the test notebook.

RFI was then radiated at the device under test, with the power slowly increased and monitored until the level of interest was reached. Any peculiarities in the operation of the device under test were recorded, as were the levels of RFI under which they occurred. The frequencies of the radiated signal at that point were also noted and recorded.

Conducted RFI Susceptibility Tests

If a test involved connection of a CRT terminal to a computer through EIA RS232C cables, the cables were draped in an exact pattern, which was repeated for every test of that type, so that all measurements were repeatable and all variables except the ones measured during the test were eliminated or at least "normalized."

These results were very carefully tabulated, and then plotted on graphs to provide more useful data. All test results were compiled and reduced to a chart showing the frequency, level, angle of radiation, and any other pertinent details about the tests. From these test results, conclusions could be drawn that were reliable and accurate in all respects.

Conducted RFI Emissions Tests

For all emission tests, including RFI and EMI emissions, the equipment was taken to a professional FCC-approved test range operated by Radiation Technology, at 18675 Adams Court, Unit G, in Morgan Hill, California. There, the equipment was tested to Class A industrial minimum emission requirements for conducted RFI emissions.

Radiated RFI Emissions

Radiated RFI tests were also conducted at the outside FCC-approved test range at Radiation Technology. The Radiated RFI tests on the equipment tested at the time were a little surprising until the results were analyzed. There were several empty EIA RS232 "D-connectors" on the metal back panel of the computer that was being tested, as well as some empty "D" holes that did not even contain connectors. The radiated RFI tests proved that those empty connectors were radiating RFI. Even after those empty connectors were covered by FCC-approved RFI shields, we were still getting RFI radiated from the back panel through the empty "D" punch-out holes. They had to be completely covered with conductive material.

EMI Tests

EMI tests of CPU and peripherals or a computer system can be carried out in several ways. A direction-seeking compass is a very handy piece of equipment during these tests, as borne out in the Chapter 1 example of EMI. In addition, there is some very sophisticated equipment available for EMI testing. A good source of information on this equipment and its use is available from seminars held by such companies as Interference Control Technologies and KeyTek. Also, magazines that cover this subject (e.g., *EE Evaluation Engineering*) are another great source of information on the latest equipment and techniques.

Radiated EMI Susceptibility

Radiated EMI tests were carried out in our screen room. In our quest for reliable sources, it was found that there are several sources of radiated EMI that would never have been suspected under normal conditions. Some of these include fans, transformers, fluorescent light fixtures, and even air compressors and washing machines! Each of these devices was moved into the screen room (except the washing machine), and measurements made on the EMI emanating from them while in operation, with nothing else turned on except the test equipment. This formed very good background information for use in the field and in the lab.

There are more easily manipulated and accurate EMI sources for use in reliable testing. These are manufactured by such companies as KeyTek Instrument Corporation of Burlington, Massachusetts.

Conducted EMI Susceptibility

In this series of tests, the EMI was impressed on the AC line that fed power to the devices under test. It also was directed at I/O cabling and other wiring going into the devices under test.

Conducted EMI Emissions

These tests were conducted at the FCC-approved test range operated by Radiation Technology. The results were printed out by the computer at the test site. These results were then analyzed and graphs generated from it for ease of understanding.

Radiated EMI Emissions

Some radiated EMI tests were conducted at John Howard's Radiation Technology site. Others were carried out within the confines of our screen room. The tests that yielded the most usable data were those involving simple but effective setups with easily acquired equipment. Each new setup was measured for background EMI before tests were begun with powered-up equipment. The background EMI was taken into account in compiling results. This is very important for purposes of repeatability and accuracy.

Purpose of Tests

The purpose of all these tests is to ascertain whether the device under test can withstand RFI and EMI being radiated at them, and to determine (if the device is an anti-RFI/EMI device) whether the device stops RFI/EMI. The filter circuit described in the foregoing discussion of filter-input power supplies was very carefully tested for effectiveness. It was particularly important to make "before-and-after" tests. By connecting up the equipment under test and measuring response to radiated RFI and EMI without the filter, then inserting the filter into the AC line, we could measure the "insertion loss" of the filter under exact, repeatable conditions. These insertion-loss measurements at many frequencies enabled a plot of these measurements, in order to outline the exact band-pass or band-stop characteristics of the device.

As said before, this type of testing is laborious and meticulous, but it yields data that is not available in any other way. Of course, recent advances in automation have given us equipment that runs tests such as these automatically. This equipment was not available just a few years ago.

The filter described in this chapter had the best RFI and EMI rejection of any device we tested. It absolutely protected the system from anything we could throw at it.

CHAPTER 4

Grounding Requirements and Lightning

We will now examine grounding requirements and the purpose for them. Anyone who believes that ground always represents zero voltage as a return path for signal and power will be surprised to learn that ground voltages can actually go "all over the map." In fact, drifts in ground potential are fairly common. This causes no problem to a circuit, as long as the supply voltages rise and fall exactly in step with the ground. But this doesn't usually happen. If spurious signals and noise appear on the ground inside the computer (as compared to the DC voltage level), we are in for trouble!

Ground potential disturbances over fairly long periods can generally be tolerated—as long as the ground and supply voltages vary in the same direction. But short-term high-frequency and high-amplitude signals impressed upon ground, as shown in Figure 4.1, are the ones that can really do the most damage to system reliability. Figure 4.1 is a simulated oscilloscope presentation of "white noise" found on the chassis ground inside a representative desktop computer.

An explanation of the mechanism for this type of ground noise is now in order. By simplifying the problems so that we can show relationships using Ohm's Law, we can easily see that ground noise voltage amplitude is dependent on two things:

Vertical = 0.1 V/cm

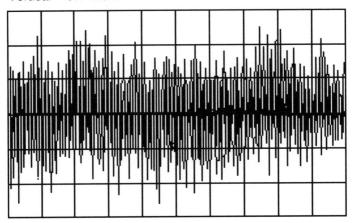

Horizontal = 0.1 µs/cm

Figure 4.1 Simulated oscilloscope presentation of ground noise.

(1) the amplitude of the current flowing through that ground, I, versus (2) the resistance (or more accurately, the surge impedance) of that ground, Z. Obviously, anything that tends to increase that ground current with a given surge impedance, Z, will increase the noise voltage amplitude, E. Conversely, anything that increases the surge impedance with a given current will increase the ground noise voltage amplitude, i.e.:

$$E = IZ$$

where E = noise voltage in volts
I = ground current
Z = surge impedance in ohms

Surge impedance of a circuit is dependent on two things: the frequency and the inductance/capacitance and resistance in the circuit. The circuit contains resistance (R), inductance (L), and capacitance (C). Surge impedance involves a second-order analysis, which is very complex due to damped oscillations and voltages. Second-order systems are those in which energy is stored in two different ways. They are most easily solved using Laplace transforms.

For an RLC series circuit, the loop equation is

$$E = L \frac{dj}{dt} + Ri + \frac{1}{C} \int_0^t i \, dt$$

It is an absolute fact that a ground can actually act as a tuned circuit at some particular frequency. Ground can become resonant at a frequency that depends on the ratio of the R, L, and C in the ground's conductors.

Strictly speaking, ground is actually a return conductor completing the circuit from a load back again to the source. This conductor has both inductance and stray capacity as well as a very small resistance. The combination of these form a series tuned circuit at a frequency that may cause it to become troublesome.

Of course, the higher the surge impedance of the ground circuit, the higher the ground voltage will exist on it and the more troublesome will be the result. As Figure 4.2 points out, the ground plane has some distributed capacitance. There is some small amount of inductance in the ground wire, as well as some resistance. All these elements add up to a circuit that will be higher in impedance, especially at the frequencies in the higher RF range.

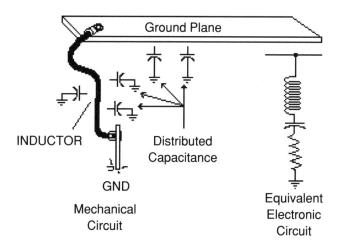

Figure 4.2 The "tuned" ground circuit.

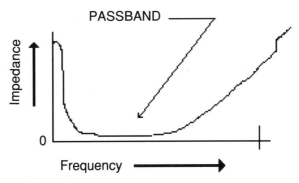

Figure 4.3 Ground resonant frequency.

Looking at the impedance-versus-frequency ratio of such a circuit, it would look somewhat like that shown in Figure 4.3. The "tuned" frequency is better known as the *resonant* or *passband* frequency. It is called a "passband" because its impedance is at its lowest value across this band of frequencies (effectively becoming a "short circuit to ground"). As can be seen, the broader the span of the passband (frequency range across which its impedance stays at a low value), the better a ground it makes. Also, the lower the impedance at that frequency range, the better.

The trick is to keep the low-end frequency of the ground's pass band *below* the AC line's frequency, and yet make the upper end frequency as high as possible (in other words, make the span of the low-impedance "notch" as wide as possible). This is easily done, of course, by increasing the cross section of the ground itself. This has the effect of significantly reducing the value of the L and R components. It also very slightly increases the value of the C component. This is the reason why properly designed grounds are always large in cross section. By going to a larger ground conductor, we are lowering the inductance and resistance of that conductor. At the same time we increase the capacitance, which lowers the frequencies at which that conductor appears resonant; i.e., the more capacitance we can exhibit in a groundplane toward actual earth ground, the lower the frequency at which the ground plane becomes resonant, thereby lowering the surge impedance of that ground.

Actually, placing parallel paths from the groundplane to earth ground is another way of doing the same thing—*if* the parallel paths all end up at *one spot* (termed the "single-point ground"). It is the same as placing other *RLC* circuits in parallel, each tuned to a slightly different frequency, because each wire would be slightly different in length, distributed capacitance, etc. Actually, this is a favorite trick of RF engineers to widen the passband of a filter.

Carry this one step further and connect the grounds from several board slots on a mother-board back plane to a single point, and you have a "tree" of grounds, all meeting at one single point. Noise-suppression-wise, this is the best of all possible worlds. With this arrangement, each signal return's own noise is not added to the noise in the one following, and so on.

This explains why different signal returns on a PC board, all connected to one point on the board, form a single-point ground that works better than a "daisy-chained" ground[1] where all grounds form a chain. In a daisy-chained ground, each ground conductor forms a series circuit that is tied to the next, etc. If all grounds are daisy-chained in series from one ground to the next, the frequency bandwidth of the passband gets progressively narrower with each added length of wire added in series.

Now, let's take the ground impedance equivalent circuit shown in Figure 4.2 one step further. Let's say that each linear foot of ground wire looks like a series capacitor and inductor, connected in series with a resistor similar to the equivalent circuit in Figure 4.2. Now hang 20 of those circuits end-to-end, in series, and connect the "top" end of this series string to our ground plane. Connect the other end to earth ground. As you can see, each "link" in the series circuit will have it own voltage drop and its own resonant frequency.

If you analyze the information in Figures 4.2 and 4.3, you will find that the lower the shunt impedance to ground, the less voltage is developed from our signal ground plane to earth ground. The idea is to keep the noise voltage present on the ground plane (due to the *developed* voltage drop across the

---

1. See Glossary.

"tuned circuit" ground wire) to as low a value as compared to earth ground as possible.

A Closer Look at Lightning

As was shown in the earlier story about the lightning strike that I witnessed in Oklahoma, the results of a nearby lightning strike can have more far-reaching consequences than just at the point of the lightning strike.

Much is known about the characteristics of a lightning strike, but not a whole lot of that knowledge has been widely circulated. After all, who cares? But let's suppose you, as an engineer, are faced with designing-out the effects of possible nearby lightning strikes. You had better know and care about lightning and its effects!

Now, let's assume that each linear foot of earth looks like a series circuit—a capacitor, inductor, and series resistor similar to the equivalent circuit in Figure 4.2. Again, lay 20 of those circuits end-to-end in series and connect one end to a power source and the other to a solid earth ground. Here again, each "link" in the series circuit will have its own voltage drop and resonant frequency. This analogy explains why earth itself has a certain voltage drop per linear foot from one foot to the next. This phenomenon is called the *ground voltage gradient* and is due to the series impedance of the soil, itself. It is naturally a very fleeting voltage (especially during a lightning strike) and would be very difficult to measure. Soil composition affects the earth's ground impedance—being generally lower, the more moisture it contains. Another factor that affects ground impedance is the amount of salts contained in the soil.

This information is presented to provide a basis for understanding other effects discussed later. A similar action takes place when current flows through air. This phenomenon is depicted in Figure 4.4, which shows both air-space and earth ground voltage gradients.

Lightning has always fascinated me, because of the horrendous amount of raw power contained in it! (But I also have the

A Closer Look at Lightning / 87

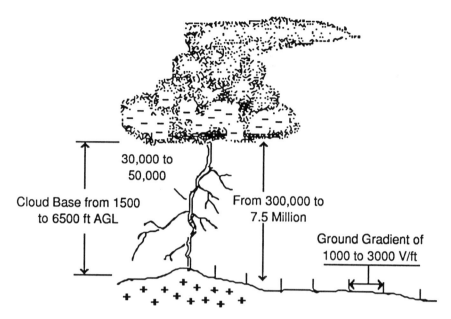

Figure 4.4 Lightning ground gradients. While we know that the voltage from cloud to ground can be upward from 300,000 to more than 70 million V, and the current in a strike can exceed 30,000–50,000 A, the voltage gradient of the ground outward from the strike is undefined. This is especially true for any single strike, because much depends on soil composition, wetness of soil, etc. The voltage gradient under certain conditions can exceed 3000 V/ft.

healthiest respect for it.) As can be seen from Figure 4.4, the values of voltages and currents that form a lightning strike are exceedingly large. The air being heated to incandescence by the tremendous amount of current contained in the bolt is what causes the flash of light and all the noise associated with a lightning bolt.

Let's take an even closer look at a lightning bolt's anatomy, which may give us further insight into the mechanism by which it generates such huge amounts of RFI, EMI, and voltage surges. Observations have been made, both by me and others, which bear out the facts that follow. More research is yet to be accomplished. A research project on lightning is currently in progress, and information gained will be made public.

During the initial phase, (Fig. 4.5), the cloud drifting over

88 / Grounding Requirements and *Lightning*

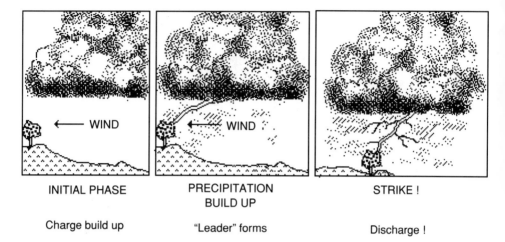

INITIAL PHASE	PRECIPITATION BUILD UP	STRIKE !
Charge build up	"Leader" forms	Discharge !

Figure 4.5 Lightning phases.

earth has accumulated a charge during its buildup and release of precipitation, etc. The mechanism existing behind this particular phenomenon of our scenario is not universally understood (or at least totally agreed on).[2] No presently known[3] theory exists that satisfactorily explains a mechanism capable of generating the tremendous amount of power contained in a lightning strike. (30,000 A times 75,000,000 V = **POWER**). However, the information that appears in this book will soon be supplemented by a research project being proposed at this writing—which if successful, may change forever the previous theories about how these tremendous charges are formed in thunderclouds. It may also explain how clouds are able to supply the huge amounts of voltage and current known to exist in a lightning strike.

At any rate, this cloud charge forces an opposing charge in the earth directly under it, acting as one plate of a very large capacitor. The opposing charge in the earth moves under the cloud—wherever it goes, following it like a "shadow." The volt-

---

2. An article covering this particular facet of our subject appeared in *Scientific American* some time back.
3. A hypothesis has been postulated by the author, which will be proven in actual field tests when the lightning study gets under way.

age difference between the charged cloud and the oppositely charged earth can become very high.[4] Moist air has a certain breakdown voltage per linear foot—which when exceeded, will cause the air between the cloud and ground to become ionized and therefore become more conductive. This breakdown-voltage value depends on several complex factors, such as altitude (or barometric pressure), humidity, temperature, dew point, and wind velocity.

Soon, as shown in Figure 4.5, an ionized column of air begins to form, extending from the cloud toward the ground. This shaft of ionized air is called a *leader*. When the leader gets within a certain distance from the ground, the matching leader[5] comes up from the ground toward the cloud to meet it. When completed, this leader exhibits a lower impedance than the damp air and presents a better path to ground. When this path's resistance drops low enough, it suddenly breaks down and "flashes over." At the point of flashover,[6] the sudden rush of current flowing through the column of ionized air heats the air to incandescence, creating the brilliant flash associated with a lightning bolt. The air expands from the instantaneous heating, then cools and contracts again, causing the crash or boom called "thunder." The tremendous amount of current flow causes a rapidly expanding magnetic field to build up and then collapse. In this respect, the cloud and ground are much like a giant capacitor and inductor. The cloud and ground form the "plates" of a giant capacitor, and the air between is the "insulator." At the point where flashover occurs, the ionized air (which is now superheated to the plasma state) then forms an inductor and some resistance.

The lightning discharge passes current through the ionized air, creating a huge magnetic field—until the voltage is "bled down" to the point where it can no longer sustain current

---

4. Various sources claim from 300,000 to 100,000,000 V. (We'll let you know, after we've measured it.)

5. Created by "corona point discharge" from a high point on the ground within the induced field.

6. The point at which a material (normally nonconductor) has ionized until its impedence is low enough to allow current flow at the voltage being applied.

through the air's impedance. The resulting current decrease causes the magnetic field that has built up to now collapse.

(At this point, I shall interject a part of a hypothesis I have postulated and has yet to be proved . . . but hopefully will be by the time you read this.) My belief at this time is that the established path from cloud to ground has the properties of an inductor, and that the collapsing magnetic field induces voltages that tend to keep current flowing, just as it would in a coil. At the extremely high currents involved, the current behaves almost as though it had a considerable amount of mechanical inertia[7]—enough, in fact, to cause current to continue to flow even after the voltage has depleted.

I also believe that at this point the *polarity of the charge* between the cloud and ground actually begins to reverse due to a phenomenon known as "overshoot."[8] Only when this charge builds high enough does the current again flow—constituting another "stroke" . . . but this time in the opposite direction. Many milliseconds (actually as much as a tenth of a second) may elapse between the several separate "strokes" of a particular lightning discharge.[9] This hypothetical reversal phenomenon is illustrated in Figure 4.6.

This theoretical oscillation may take place many times during one lightning bolt. The frequency and damping-out of this oscillation depends on many complex factors, among which are the surge impedance, and overall distance from cloud-to-ground. (As a matter of fact, the distance may be increasing due to drift of the cloud in the wind, etc., which increases the impedance and lowers the current. At the same time, the frequency will probably decrease during the damping-out of the current oscillation.[10]) This multiple-discharge current accounts for the "flickering" of the lightning flash. As stated in the footnotes, this was proved recently by a video-taped sequence, shot during an actual tornado/thunderstorm at night near the Colorado

---

7. This phenomenon is sometimes called the "flywheel effect" of a coil.
8. See Glossary.
9. This has recently been proved by a video sequence shot by the author at night, and single-framed to show as many as eight separate strokes during one lightning strike.
10. Also proved, in the same manner.

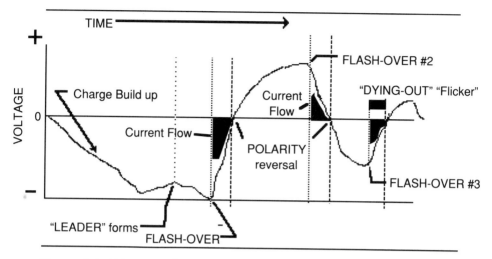

Figure 4.6 Voltage waveform of a lightning strike. The waveform shown is the author's rendition of the approximate voltage waveshape required to explain the multiple discharges found in a large percentage of lightning discharges. In this graph, current is flowing (and therefore luminosity occurs) only in the time slices containing the black portion of the waveshape. Notice the extremely fast rise-time of the current.

River, which shows as many as eight separate luminous periods during a single lightning strike lasting less than half a second. A copy of this video is available for viewing, by special arrangement.

The uneven current and multiple-discharge oscillation of the lightning discharge creates radio-frequency (RF) noise and a varying magnetic field (EMI) of multiple frequencies. The high voltages and currents radiate RF signals for miles, creating the "static" commonly heard on AM radio stations. These radiated frequencies cover the entire radio spectrum, creating white noise (a very large source of RFI). At the same time, the immense alternating currents build up tremendous alternating magnetic fields, which cut across any conductors that might be nearby and induce alternating surge voltages in them. This is the cause of the EMI that results from a lightning strike.

The alternating current flow through the ground from the strike sets up alternating ground gradient voltages, creating ground noise for any circuit completed by it, for several hundred feet in every direction. The ground gradient can exceed several thousand volts per linear foot of earth.

Combating Ground Noise: RS232C versus RS422

To minimize the effects of ground noise, some peripheral and computer manufacturers have gone to a more reliable mode of signal transmission known as "EIA RS422." As shown in the Figure 4.7, the main difference between EIA RS232C and EIA RS422 is that while RS232 and its variations rely on one "hot" wire and a ground, RS422 uses two "balanced" outputs, neither of which are grounded or even referenced to ground. Of course, this is not the only nor even the most important difference between the two. Voltage levels representing a high ("1") and a low ("0") are different. This, coupled with the large threshold, allows much longer transmission cable lengths. Whereas RS232 is specified to be reliable at cable lengths of 50 ft or less, RS422 can be reliable at over 1000 ft.

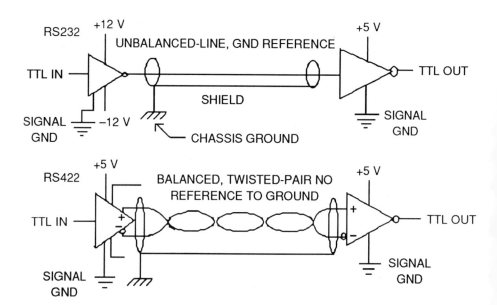

Figure 4.7 RS232 versus RS422 transmission.

Since neither of the balanced outputs of the RS422 driver is grounded, variations in ground level due to gradients and other disturbances have no effect on the signal. Therefore, signal reliability is no longer so closely tied to ground potentials.

Other Considerations

Let's take a close look at analog-versus-digital logic grounds. In the case of a CRT terminal, there are both analog and digital circuits required in order to arrive at the end result. But mixing the grounds of these two technologies can be disastrous, and usually is.

What is the problem with this arrangement? The very term "analog" indicates a *varying* voltage, rather than the "1s" and "0s" of the digital world. Any varying analog voltage causes varying (or even *alternating*) currents. Alternating or varying currents flowing through wires are good sources of radiated radio frequency interference. Carrying this a step further, a ground for alternating current circuitry must, by definition, carry alternating or varying current. This, in turn, causes a varying ground voltage by the same mechanism outlined earlier. If binary logic uses the same return path, these varying ground voltages or noise will be impressed on the digital signals. Thus ground noise will affect the logic if the noise voltages reach the threshold required to create a digital data bit. If the ground noise level doesn't quite reach the threshold, it can elevate the ground voltage to the point where only a very small increase from anywhere else will surpass the threshold. Radiated EMI from a flyback high-voltage transformer and associated circuitry becomes conducted RFI when the EMI "cuts" nearby wiring and can add to the effects of ground noise to put the logic over the threshold.

The points discussed should begin to make clear why mixing analog and digital grounds is not a good idea.

Lets suppose, for example, we have the situation mentioned in Chapter 1 where pins 1 and 7 of the RS232 cable were connected to chassis ground as well as logic ground. This automatically ties the two grounds together at a very critical point, where incoming, already noisy signals are arriving from a remote terminal. If the threshold of the receiver is not correctly set up, this additional noise will mix with the signal and create "garbage" signals. The least this will do is change characters by changing the state of bits, and it may even create "garbage" characters of its own. This reduces the reliability of transmitted and received information to the point that it may affect the entire system. In fact, in one case I remember it caused the computer to respond to "ghost" commands and do all sorts of weird things. This is especially true where single-strike key commands are used to generate software actions in the computer.

External-to-System Grounds

One of the worst ground problems you (the engineer) may ever have to deal with is one *external* to the computer you might be working on—the *facility* ground. Even though it is obviously beyond the control of the system or board-level design engineer, you should be aware of all facets of this problem so that you are armed with the knowledge needed to combat it.

There are several wrong ways to connect power to a computer in an installation site, but scarcely more than one right way. There are three common wiring schemes you will find at a typical customer's computer installation site. They range from fair to bad.

Figure 4.8 shows a very poor way to connect a computer

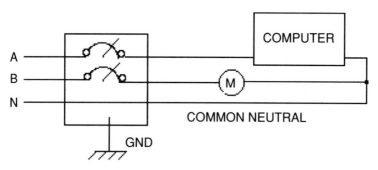

Figure 4.8 Poor AC power technique.

system, and we shall now investigate why it is so poor. In this example, the customer has two circuits (A and B) connected through the same main breaker panel. Circuit A is connected to a computer. Circuit B from the same source is connected to a very large AC motor (marked "M" in the sketch), *which shares the same common neutral wire as the computer.*

What's wrong with this scheme? Suppose we "bring up" the computer on circuit A and get it running. The computer has established a certain amount of current through the common neutral after the initial inrush current surge is over. Then suppose further that, with this "quiescent" current having settled to a steady state, someone suddenly turns on the motor (M), on circuit B. We suddenly get an inrush current surge to the motor (M) from circuit B, which is conducted *back to the source through the same return path as the computer current.* This causes a huge increase in the current through the neutral wire, which increases the voltage drop across the neutral wire. This voltage drop is subtracted from the input voltage, and is felt at the computer as a lowered voltage across it. What we have here is an *induced sag* in the applied voltage as a result of turning on the motor. At some point, the accelerating motor will reach its operating speed and the current surge will give way to a constant current level. The input voltage across the computer will then rise, causing what appears to the computer to be a surge.

Figure 4.9 depicts a better way to connect the computer into an existing AC power wiring setup. But this scheme also has many problems. Even though the two components have sepa-

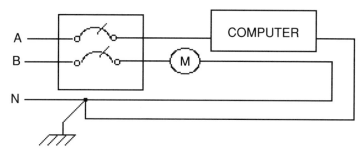

Figure 4.9 Better power grounding.

rate neutral wires, they still join together at the ground and connect to a single conductor going back to the main neutral. The fact that they are joined at a single-point ground helps considerably. But we are still not there yet.

Figure 4.10 shows the best way to connect these two completely different loads. Here the main power going to the panels supplying the two loads comes from a separate source. The neutrals are separate wires all the way back to the source, at which point they join at a common single-point ground. This method of supplying power to the computer is known as a *dedicated* AC line and an *isolated* ground. This is an absolute requirement when installing the average small computer in order to reduce sags, surges, and noise on the ground. Otherwise, noise and power sags and surges from other equipment will eternally cause computer problems.

While we have spoken of EMI and RFI, there is one other source of problems we have not talked about. This is ESD, or electrostatic discharge. We will now take a close look at the mechanism through which ESD operates.

ESD

ESD is a small-scale lightning generator. The only difference is the voltages and currents involved are on a much smaller scale. The fact that the current is minute and the proximity of the charged body is hundreds of times closer means that the voltage required to flash over and create a spark is much lower.

The mechanism by which ESD operates is very easily understood. As you probably already know, when two insulating materials come into close contact (such as a rubber shoe sole against a carpet), in brushing by, they pass electrons from one to the other, creating a difference in potential. If this process continues, the voltages build up. If there is not a path of sufficiently low impedance to allow this charge to "bleed off," it will continue to accumulate until the voltages become very high. If, during this process, your body comes near enough to a low-impedance path to ground, flash-over occurs, and a spark takes place. *ZZAAP!* This is the familiar shock you get when

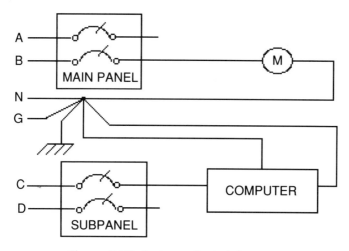

Figure 4.10 Best grounding technique.

putting your key in the lock after walking down a carpeted hallway on a cool, dry day. This spark can be very destructive to computer parts, especially if your computer is the path to ground. CMOS technology parts are particularly susceptible to this destructive phenomenon. Another different, but very destructive phenomenon is . . . EMI.

An Example of EMI

A few years ago, I was a member of an engineering team that was designing and implementing a microcomputer system for slot machines. The purpose of this computer was to count several events, such as the number of coins inserted, the number of times the handle was pulled, the number of coins paid out, and so on.

Now, this was theoretically fairly easy to do, but several factors kept getting in the way. For instance, the interior of a slot machine is the world's *worst* environment for a microcomputer! The slot machine is an electromechanical nightmare. Sliding electrical contacts that tend to arc, magnetic solenoids (which are terrible alternating magnetic-field generators), magnets, etc.—all the "ghoulie" computer-killers are found in slot machines.

That was not the worst. After we had found and fixed most of those problems, and had even begun installation of our product onto actual machines into gambling casinos, we suddenly came face-to-face with . . . ESD!

All manufacturers live in mortal fear of an unforeseen, but very real, threat that is totally out of their control—a user habit that causes a system failure. That type of problem is very hard to design out if you have no idea it is going to take place.

Our microcomputers were housed in a plastic case. One end of the case was a metal heat sink for the power supply. The cases were attached to the side of the slot machine, heat sink at the top. A red digital display faced the customer.

This is the situation we found ourselves faced with:

1. Suddenly, our computers were "dying" left and right.
2. Every failure could be traced to a failed clock chip.

At first, we suspected a bad run of clock chips from the manufacturer. But after extensive tests, we found that the same chips we were installing in the machines worked in the lab without failures. What could be going on in the gambling casino that we were not doing in the lab?

So we investigated. We watched some of the machines at the casino, trying to get a clue as to what was taking place. It was noticed that customers were using the microcomputer's case for a place to put their drinks, while they played the slot machine.

Meanwhile, I had disassembled one of the failed micros that had been brought back to the lab. As usual, the clock chip had failed. I noticed the heat sink was a "T"-shaped aluminum extrusion, and the PC board was bolted to the center of this "T." As I got further into the computer, I noticed that the +5-V track from power crossed under the center "T" on the heat sink. Then it hit me—ESD! When a customer walked across the carpeted floor, he or she picked up a charge from the carpet. When he or she set their drink up on the heat sink of our microcomputer, they were literally charged up. As their hand came near or

An Example of EMI / 99

contacted the heat sink, there was a discharge path to ground through the heat sink! Enter our old friend "zzaap"!

To fix this problem presented a real challenge. We needed to "harden" the computer against ESD. Can you imagine what changes were required?

The Fix

First of all, to test the ESD theory, I needed a "zzaap tool" that would generate a very high-voltage spark toward nearly anything I held the tip close to. So I built one. Sure enough, every time I struck a spark to the heat sink, it "zzaaped" out the clock chip on the computer. See Figure 4.11 for a diagram of the "zzaaping tool" I built for this test.

There are several commercial tools for this type of testing now available. I used a "zzaaping device" about 2 years later, and found it much more flexible and safer to use than my "zzaap tool." It worked very well, putting out an adjustable voltage pulse. The pulses had a very fast rise-time, followed by a damped oscillation, and covered every frequency of interest. Several different antennas can be used with this particular "zzaaping" device, allowing either radiated or conducted interference testing.

Once we had the "zzaap tool", here's what we had to do to lick ESD's "zzaaps" in the slot machine computers:

1. First, after surveying the situation and very closely inspecting the area where the heat sink and the PC board

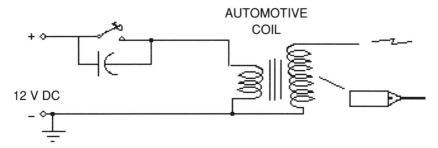

Figure 4.11 "Zzaaping tool."

met, I could see that the heat sink was attached directly over the +5 V line on the PC board.
2. The PC board layout was such that the only path to ground for a high-voltage discharge into the heat sink was through the +5 V track to the clock chip. If we were to cut an inch off the center of the heat sink, it would completely clear any tracks on the board. So I had an inch of the heat sink removed. Also, a better, lower impedance path to ground for the heat sink was required. I provided that low-impedance ground by using a 6-in. piece of $\frac{3}{8}$-in.-wide tinned-copper braid.
3. A high-voltage capacitor [≈0.05 MF (microfarads or so) at 600 WV (working volts)] was connected between the +5 V track and the new ground. A 100-V MOV was connected paralleling this capacitor from +5 V to ground (of dubious value, but added insurance).

After these improvements had been made, I could zzaap the heat sink with as long and fat a spark as I could possibly generate and the clock chip never blinked! I beefed up the "zzaaping tool" for even higher voltage. Even with a 4-in. fat, blue spark, the computer could stand all the ESD I could throw at it. We thoroughly tested this modification, then installed the change on every one of them. We never had another clock chip failure after that. The moral to this story is, however, that the ESD-proofing should have been designed in at the start.

Now let's take a theoretical device and do an ESD hardening job on the design before it leaves the engineering area. At the power line input to the computer, we put a filter input as discussed in Chapter 2. Then we make certain that no more than a 2-in. length of power cord is left between the power entry and the power supply itself. Next, the power supply is completely shielded, and built to a rating of at least 200% the required current-carrying capability.

The next part to be attacked is the grounding scheme for the entire unit. A single-point ground, with very-large-gauge bonding strap, is strung from every piece of sheet metal to the single-point ground that joins the AC green-wire ground at the AC input. This is done even if the sheet metal pieces are in contact with each other. (Who knows, down the line during assembly, the sheet metal may turn out to be anodized or some other coating, which will make it relatively nonconductive.)

EMI and RFI shielding of the entire unit must be considered. It should encompass the entire unit, with any and all parts that have digital or analog signals completely shielded from each other and from the outside world. This EMI/RFI shielding must in turn be connected to the single-point ground by a separate bond to each shield.

Regardless of whether your design embodies a mother board or back plane, the following steps are necessary:

1. A separate large-gauge wire from each supply voltage to the entry point on the mother board or back plane. (Wire gauge should be taken from the wire-gauge charts, which show resistance and current-carrying capacity, as well as other information such as number of strands, size of each strand, etc.) Manufacturers of wire and cable supply these charts.
2. A power supply ground return wire or wires that total to at least twice the cross section of all the "hot" wires combined. (This is to prevent high ground impedance with accompanying ground noise.)
3. Each individual feed wire onto the back plane or mother board should have its own lossy toroidal core on which at least six turns of the feed wire have been wrapped.
4. Large-capacity capacitors should be connected between each of the feed wire entry points and the ground plane.
5. The ground plane should consist of a complete sheet of copper clad across the bottom of the board, with appropriate openings to prevent short circuits. This ground plane should be tied from one point only, with wire that surpasses the cross-sectional requirements mentioned above, to the single-point ground.
6. Any tracks on the mother board or back plane that carry power must be at least twice the cross-sectional area required to carry the maximum expected current.
7. The number of connector pins that carry ground onto a

daughter board or "plug-in board" must be sufficient to provide power at as near a zero voltage drop as possible, and in no case a drop of more than 0.5 V.
8. Power-carrying pins should appear on *both* sides of the connector, opposite each other, and at the opposite end of the connector from the grounds.
9. Do not depend on sheet metal to carry either power or ground—use conducting wires or braid.
10. Make all grounds and power feed wires *as short as possible*.

Follow the above guidelines for the least problems possible with the power and ground.

CHAPTER 5

Noise Susceptibility and Emission

As we previously established, "noise" is a common term used to refer to EMI, RFI, and power disturbances in general. We also determined that electrical noise propagated *into* a computer (either conducted via AC power cord or radiated from the air) is known as *noise susceptibility* (see Fig. 5.1).

Put another way, noise entering a computer from the outside and then causing problems is termed as *noise susceptibility* on the part of the computer. Conversely, noise being propagated *out* of the computer that it was generated in is called *noise emission* (see Fig. 5.2). Therefore, noise being either radiated or conducted from inside the computer out to the outside world, is called noise emission.

As can be seen from Figures 5.1 and 5.2, noise can follow either of two paths: (1) conducted by way of the AC line cord (or other connecting cable) or (2) by being radiated through the air. Notice I said "conducted or radiated"—because noise can take either of two avenues . . . *conducted* or *radiated*. If radiated, the noise is traveling through the air like radio waves as shown in Figure 5.3.

If conducted, the noise is going by way of the AC cord or other wires as shown in Figure 5.4.

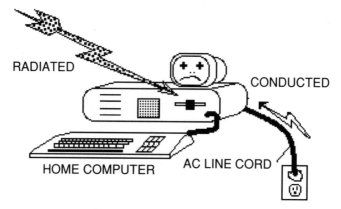

Figure 5.1 Noise susceptibility.

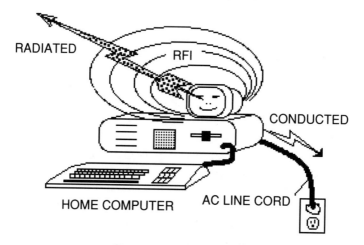

Figure 5.2 Noise emission.

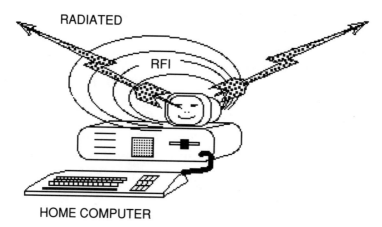

Figure 5.3 Radiated noise.

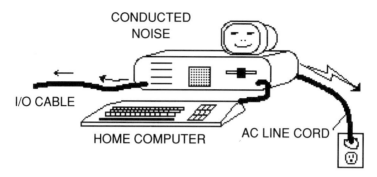

Figure 5.4 Conducted emissions.

Now we are about to discover that the terms "noise susceptibility" and "noise emission" really refer to the same phenomenon, and are different only in the *direction* they are propagated. This may well be a new concept to the reader: *any path for EMI, RFI, and other noise is a two-way street.*

Now, let's imagine a hypothetical case where we have a computer that has been proven to be guilty of illegal RFI emissions.[1] If the supposition that "the tendency for noise emission is the alter-ego of the tendency to be noise susceptible" is correct, then a computer with an illegal emission level should also be susceptible to the same RFI frequencies from the environment outside the computer—correct? And, indeed, it was, in repeatable engineering lab tests. Think about it . . . what's so different about the *direction* the noise takes, inside to outside, or outside to inside? If there is an RF leak, there's a leak. There are generally no direction-sensitive parts in the leak path to prevent noise currents from flowing in *either* direction.

Most band-pass filters are designed to match a specific input and output impedance. Therefore, given that the impedances are matched at both ends, a series band-pass filter will propagate *only the frequency at which it has the lowest attenuation.*

---

1. It would not meet requirements of Part 15, Subpart J of the FCC Rules and Regulations governing emissions.

The use of either term—noise emission or noise susceptibility— is simply another way of saying that our computer's noise filtering and shielding is faulty . . . in *both directions.*

Noise susceptibility and noise emissions are alter-egos of the same problem.

How do we know this? In controlled laboratory tests, it was proved that the noise frequencies at which the computer or peripherals measured the highest in emissions were the *same* frequencies the computer or peripheral were most susceptible to from outside. By running both sets of tests in the same area with the same array of equipment, and as near as possible the same general setup, we definitely proved there was a correlation.

It follows, then, that susceptibility to inbound noise can result from noise taking the same path that it would have taken had it been outbound noise emissions. Therefore, the fact that there are noise emissions at some particular frequency means that there may also be a noise susceptibility problem at that same frequency.

Said another way, *noise emissions are the outward manifestation of a problem that can also cause noise susceptibility.*

The Theory of "Duality"

Let's go a step further, and postulate a hypothesis, which we will dub the "Theory of Duality":

A particular type and frequency of noise that is presented a path *into* a computer from outside; can also use that same path *out* of the computer from inside [assuming a noise of that particular type and frequency was generated inside by the computer].

IMPORTANT: This concept must be fully understood at this point, before we continue. For what we are seeing here is a profound manifestation of the well-understood laws of physics governing electricity. Figure 5.5 illustrates this quality of "dual paths"—both in and out of the computer. These two paths, again, are conducted and radiated.

Look at it this way . . . a *noise path* can be defined as a

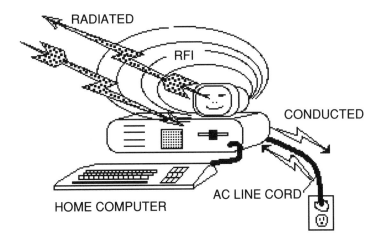

Figure 5.5 RFI duality.

"tuned circuit"[2] through which a particular frequency or band of frequencies will be propagated with the least amount of attenuation. The input and output impedances at each end must be perfectly matched no matter which way our "signal" is propagated. In the case of a noise path, this is inherently true since this filter is a "natural band-pass filter" at this particular frequency.

For purposes of definition, as far as the FCC is concerned, emissions are just that . . . EMI and RFI being emitted by the computer itself. On the other hand, noise susceptibility is not an FCC concern for the most part, because they assume that if emissions are controlled, susceptibility should not be a problem—at least, not their problem.

Noise Paths

As it turns out, there is a very direct path that noise can take before it even gets to the power supply. How about the AC power cable itself? An unshielded length of power cable makes

---

2. A tuned circuit here means a circuit that displays an attenuation curve with the point of least attenuation at some particular frequency (also known as a *band-pass curve*).

a very effective antenna. If there exists even as little as 4 in. of unshielded power cable inside a perfectly sealed metal box, RF can be induced into it and then be conducted by way of the AC cable—in either direction. And in the process, the power cable can become either a radiating or receiving antenna! RFI of the proper frequencies generated inside the box will take the "conducted" route via the AC power cable to a ground outside (as shown in Fig. 5.4). Or using the same mechanism, RFI from the outside environment can be conducted to sensitive components inside the box (see Fig. 5.1).

And what's more, believe it or not, a very small opening (such as an empty connector mounting hole) in an enclosure is all it takes for that small opening to become a "slot antenna" and radiate internally generated radio frequencies—as RFI—to the outside world. This has been proved many times in tests conducted at outdoor FCC-approved RFI ranges.

EMI and RFI Sources

A switching power supply is an excellent source of both EMI and RFI—at frequencies from 10 kHz upward to 10 GHz. Any wires or other signal paths leading outside from the switching power supply's enclosure is a prime path for conducted RFI. For this reason a switching power supply should always be well shielded, by a well-grounded enclosure of its own. It should also be provided with filters for every wire connecting to it.

There are penalties ahead for the computer that is guilty of either noise susceptibility or emission of noise. If the computer has "high emission levels," the FCC will be after the manufacturer for broadcasting without a license! (Seriously, the FCC may cite the manufacturer for not meeting the maximum-emission requirements of Part 15, subpart J of the communications rules relating to radio noise emissions.) This is mostly because the computer will interfere with radios, TV sets, and any other kinds of communication devices that depend on "broadcasted" signals. On the other hand, noise susceptibility will be a constant source of system problems in a poor environment.

In other words, in saying the computer has a "high noise

susceptibility," we are saying it would be very sensitive to environmental noise—both conducted and radiated. Conversely, we are also saying it may be guilty of emissions at the same frequencies.

Case in Point

One company had sold several hundred computer systems that were so noise-susceptible that the hardware systems engineer (whose job it was to solve all the company's field-service problems) had to temporarily issue the following warning by way of a Field Service Bulletin, which appears in Figure 5.6.

The company that issued the Field Service Bulletin was a "value-added reseller" (VAR) for the computer hardware they were selling and installing in the field. (i.e., this company was another "systems house-turned-integrator").

Being a "software systems house-turned-integrator" meant that there were very few people in the company who actually understood hardware and the responsibilities that it brought with it. They thought you could just hook it up and it would play. Very few computers are really "plug and play." They usually require considerable planning, forethought, and site preparation in order to operate in a totally trouble-free manner. Anyone who doubts this will one day pay the price.

In fact, a good friend of mine coined the phrase: "I/O problems" means "ignorant operator problems." Very well put. This saying is all too true, too many times.

Fighting Back

But take heart, there are modules and components available "off-the-shelf" that are designed to make the hardware-systems design engineer's job a lot easier. Some of those components are explored later.

It happens that, as a result of extensive research on the part of EMI/RFI equipment manufacturers, there are components available whose specific purpose is to prevent noise emission *or* noise susceptibility. Although some of these products might be

FIELD SERVICE BULLETIN

List of Equipment Incompatible with Computer Systems

The following list of equipment includes equipment likely to be at a user's installation, that has been found to be *incompatible* with the average small-business computer equipment. Any equipment on this list should be *separated from,* and by all means *should not share AC power with* the average small computer system in any way!

This list is probably not complete, but should give a good idea of what types of equipment are electrically "noisy" and therefore cause trouble with the average home or small-business computer. The type of equipment on this list should be avoided wherever possible, both from an environmental standpoint (don't put an average small computer near them), and from an AC power standpoint (don't connect any of these to an outlet to be shared by computer equipment).

Furthermore, if computer problems are traced to and proven to be a result of any of the equipment on this list, the customer *must* be told to make changes to remove the offending equipment to another less offensive location.

Incompatible equipment list:

- Air compressors
- Battery chargers
- Cash registers
- Calculators that print
- Power transformers
- Drill motors
- Exhaust fans
- Grinders (valve or bench)
- Heaters, electrical
- Large transformers
- Machine-shop equipment
- Public-address systems
- Reamers
- Telephones—Ringer type
- Vacuum-supply pumps
- Very-high power motors
- X-ray equipment
- Microfiche reader
- Air conditioners
- Brake-drum grinders
- Candy machines
- Coils
- Drill presses
- Electric Typewriters
- Fluorescent lights
- Golf Carts
- Ignition analyzers
- Microwave ovens
- Motors driven from AC
- Radio transmitters
- Roll-up doors
- Telephone switching
- Valve-grinding
- Welders, electric arc
- Xenon lamps
- Mercury vapor lighting

Check your customer sites for these incompatibilities!

Figure 5.6 Field service bulletin.

effective in both directions, they are usually specifically designed to be effective in one path.[3] They are usually designed for preventing *radiated noise* or specifically for *conducted noise* . . . generally not both.

---

3. "Path" here means the physical path followed by the noise (either conducted by wire or radiated through the air).

Available Equipment

Now we will examine some of the components designed for reducing or preventing RFI, and give you some data pertaining to their specifications and use. We will begin at the AC power entry into the computer housing, with the connection of the power cord.

Available Equipment 1—Power-Entry Modules

Power-entry modules are off-the-shelf components specifically designed for minimizing the RFI and EMI emissions and susceptibility of the equipment they are designed into. The manufacturer does this by including EMI and RFI filters in the module. Some of the types available even include integral fuseholders, and all are made of component plastic that is resistant to both heat and leakage currents.

We obtained samples of some power-entry modules from one manufacturer for purposes of this book, and found many interesting variations in the different modules available from just this one manufacturer alone. All sample power entry modules contain an integral three-blade male receptacle that allows removal of the power cord from the equipment. The power cord itself terminates at the power-entry module end, in a rectangular female cord end plug containing three flat-blade sockets. This female cord end plugs into the male receptacle of the power entry module. The other end terminates in a plug compatible with the normal three-hole wall outlet.

As noted, several different styles and sizes of power entry modules were obtained from Schurter, Inc., the generous manufacturer for purposes of this book. The smallest (and probably the least expensive) power-entry module is one that includes the ubiquitous three-blade AC power cord adapter combined with an EMI filter. It is specified as a type KFZ, and is rated at 1 A. The factory specs for this entry module show a noise-attenuation insertion loss of about -50 ± 5 dB, from about 200 kHz to around 100 MHz! This is very good for this small a device. For those who might be wondering, Figure 5.7 graphically illustrates the "insertion-loss"[4] profile of this filter, in the 1-A asymmetrical model:

---

4. "Insertion loss" is the ratio of power in a given signal at the input to a filter versus the power remaining at the output of the same filter, expressed in decibels (dB).

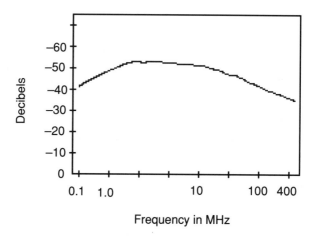

Figure 5.7 Insertion loss for type KFZ filter. Information supplied by Schurter, Inc. Graph by author.

The graph in Figure 5.7 shows the "insertion loss" expressed in decibels versus frequency from 100 kHz to 100 MHz, indicating the nearly flat response of the RFI filters claimed by the manufacturer for these EMI and RFI-filtered power entry modules. (Data supplied by the manufacturer—graph drawn by author.)

Another slightly larger module, depicted in the first photo (Fig. 5.8), is designated as a type KFA and contains a fuse drawer that allows very simple withdrawal of the fuse. This module is available either with or without a voltage-selector insert, which allows selection of one of three operating voltages (115, 220, or 240 V AC). The simple action of rotating the voltage-selector insert (which forms one end of the fuseholder) to the proper position so that it displays the selected voltage through a small window in the rear of the unit, automatically sets the device for the proper fuse length and input voltage. Nothing could be simpler or more convenient.

A second power-entry module, type KFB (not shown), is available with an "on–off" SPST switch built in, which allows power switching right at the power-entry module. Both the type KFA and KFB EMI filters maintain better than −55-dB noise attenuation from 1 to 100 MHz!

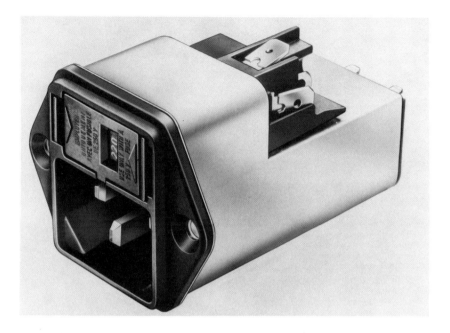

Figure 5.8 Type KFA line filter/AC entry module. (Photo courtesy of Schurter, Inc.)

The type KFC appearing in Figure 5.9 includes a single- or double-pole fuseholder, and is specified to have a −50-dB attenuation ± 5 dB, from 1 to 100 MHz.

The largest of the samples was a type CD. This module is the "elite" of the line, and includes a line filter with AC connector, an "on–off" line switch, voltage selection, and interchangeable fusedrawer for $\frac{1}{4} \times 1\frac{1}{4}$-in. (or 5 × 20 mm) fuses. This power-entry module is shown in Figure 5.10.

We wish to express our gratitude to both Schurter, Inc., P. O. Box 750158 in Petaluma, California (the manufacturer) and Quail Electronics, Inc. of Scotts Valley, California (the distributor) for the above photographs and other information.

Available Equipment 2—RFI/EMI Filtered Switching Power Supplies

The manufacturers of many different switching power supplies offer RFI and EMI filtering built into the products they

Figure 5.9 Type KFC power-entry module. (Photo courtesy of Schurter, Inc.)

manufacture. There are too many different kinds to list here, but please bear in mind that they are available . . . in almost every size and capacity you might have occasion to require. A good deal of very useful information is available from the power supply manufacturers simply for the asking, and may be obtained by circling the correct number on a "bingo card"[5] from an ad in any one of several magazines that serve this industry.

At any rate, please be aware that a complete array of power supplies with built-in RFI and EMI suppression/filtering are available "off-the-shelf," and specifications can be obtained directly from the manufacturer. Simply find an ad for this type of power supply in one of the trade magazines or any others that serve our industry and circle the number that corresponds to the ad. A list of some of these magazines appears later in this chapter.

Many of these manufacturers have written material available that thoroughly explains (in even more detail than I go into

---

5. "Bingo card" is a nickname used for those information-retrieval cards contained in most of the electronics magazines.

Figure 5.10 Type CD power-entry module. (Photo courtesy of Schurter, Inc.)

here) the different problems and the product(s) they manufacture to deal with it. An example would be brochures and handbooks distributed by General Electric on their line of MOVs, the *Transient Voltage Suppression Manual*, or *Staco's Regulator Catalog*, which includes a considerable amount of information about power-line problems and their solution to them. Another example would be Coilcraft's booklet "Data Line Filtering." Nearly every manufacturer of line filters and conditioners, UPS devices, and the like have extensive information available in pamphlets and booklets, simply for the asking. Get abreast... send off for them. Be especially sure to obtain the book on Transient Eliminators available from Advanced Protection Technologies, at 14088 Icot Boulevard in Clearwater, Florida 34620.

But be careful about being misled by generalities and indirect inferences in the magazine ads, rather than hard facts and specifications—just as in the case of any advertised product. The manufacturers of these products try to offer the most "bang for the buck," but they are in business to sell their product. The best means of comparison is to obtain a sample from more than one source, and do your own testing. Compare part A with part B from different manufacturers. In this way, you'll not only be getting the true story about the product with your application in mind, but you'll be learning more about the subject in general.

Later in this chapter a list of publications appears that you can subscribe to (mostly free) and which are rich sources of information about designs, circuits, and components.

Available Equipment 3—RFI/EMI Enclosures

Other equipment available for use in fighting EMI and RFI includes a class of enclosures manufactured specifically for this purpose. These enclosures are sold by several sources, and sometimes marketed as "TEMPEST"[6]-qualified. I shall make no specific recommendations as to manufacturers or whether one brand is better than another, because the purpose of this book is not to be a source-book for these things. But off-hand, several names such as Zero, Budd, Stantron, and others come to mind.

Most of these enclosures do the job with the aid of such accessories as:

- EMI/RFI gaskets (both the springy-braid variety and the conductive-rubber type)
- Copper–beryllium alloy spring-door-closure "gaskets"
- Various other sealing, hinges, and fittings that are made specifically for EMI/RFI enclosure purposes.

Available Equipment 4—Shielding

There are several types of shielding available for both EMI and RFI-proofing a system. Magnetic, and even EMI shielding is

---

6. "TEMPEST" is a military communications-security classification for an emission-proof environment.

easily done using "Mu-metal" shields. This type of shield is a very soft, ductile metal that can be formed into whatever shape might be required. It is a bright, "white" metal that does not tarnish. Since Mu-metal is so malleable, it can be shaped so as to completely surround a section of circuitry that would be either susceptible to or an emitter of EMI, without it. It is however, rather expensive.

Another popular shield material is copper sheet—which is used in winding certain transformers. This metal is used to form a "Faraday" shield against cross-talk and core-propagated noise by placing it between windings in many different types of transformers . . . such as power step-down transformers, power-ground isolation transformers, audio transformers, and speaker output transformers—especially those used in computers and sound equipment.

Still another type of shielding is that listed above under EMI/RFI gaskets—copper–beryllium alloy spring shields, which form a conductive but crush-proof, springy edge for doors, panels, and other removable parts of enclosures and cabinetry.

Available Equipment 5—Conductive materials

Many forms of conductive materials are available for use in prevention of both RFI emissions and susceptibility. For example, conductive-plastic caps are available for closure of unused D-shaped RS232 connector holes. Conductive rubber or plastic is available for many purposes; such as the conductive foam used for storing IC's, as mats for prevention of ESD, as well as simply as gasket material to be used at the edges of movable surfaces (and some of them are even coated on the gripping surface with adhesives).

Other Sources of Information

There are many other sources of information on this and related subjects, including a large selection of very useful free-subscription industry-oriented magazines, which can be had by qualified subscribers at no cost, simply by mailing in a free-subscription application form to the publisher. These forms are found in nearly every issue of these magazines.

A list of these magazines would include, among many others:

Compliance Engineering—published by Compliance Engineering

EDN—a magazine devoted to electronic design, published by Cahners Publishing Company

EDN News—a newspaper published by Cahners Publishing Company

Electronics—a design engineering magazine, published by VNU Business Publications

Electronic Design—another VNU Business Publication

Electronic Engineering Times—A newspaper for engineers and technical management, published by CMP Publications, Inc.

Electronic Products—published by Hearst Business Publications

EE/Evaluation Engineering—published by Vernor Nelson Associates

Test & Measurement World—A Cahners Publication

As indicated, these magazines are available by free subscription to all those engaged in this line of work. I urge you to avail yourself of as many of these free magazine subscriptions as you can wangle subscription forms for. A subscription form can be obtained from any issue that usually appears in the reception area where you work. These magazines are all "worth their weight in gold" when it comes to information about the subject of this book and other related information about our chosen profession. They offer an ongoing source of knowledge about the subject. They are our main means of keeping up with the technology in this field. I suggest that you subscribe to all these magazines, then renew only those subscriptions which you have determined are in your line of interest.

The above should indicate that there are a seemingly endless list of materials, components, and equipment available "off-the-shelf" for use in the never-ending battle against EMI, RFI, and ESD. There is also a wealth of written material available (as mentioned) in the form of product pamphlets and brochures.

These materials and components do many different and varied jobs, but all are designed for one purpose—that of preventing "Zzaaps." You . . . as the design engineer or eventual user or whatever, have access to these wondrous devices.

If you have a product that is being bothered by any of the terrible disaster-producing noise sources discussed in this book, get more information on this subject . . . or better yet, contact a consultant who already "knows the ropes."

There are many very knowledgeable and competent experts in this field who are available for consultation simply by giving them a call. They are consultants simply because in this field, they have expertise that takes years to accumulate.

This book will help to acquaint you with the basics and give you an understanding of the underlying causes of zzaaps, but it will not make you an expert. Without specific experience and training, even though you have amassed a thorough understanding of the mechanisms and sources of computer system zzaaps, you may find yourself overwhelmed. The actual causes are sometimes not easy to find. But not to worry . . .

If you should find yourself in the position of needing expert help, don't hesitate—simply contact a competent electromagnetic-compatibility consultant who specializes in ESD/RFI/EMI. Then sit back and relax, and let the expert solve your problems. Sure, it will be expensive for a short time—but it may end your problems . . . hopefully **FOREVER!**

CHAPTER

6

Other Reliability Factors

In this chapter, we shall consider some other factors (besides those we have already discussed in foregoing chapters) that can affect system reliability—those factors over which the computer manufacturer has little or no control. Included here will be some factors that might not normally be considered as an "exclusion" (for warranty purposes) as a cause of failure. These factors have a direct bearing on the reliability of a computer system if their effects are not "designed out." Yet most of them are influences that, as we said above, cannot be directly controlled by the manufacturer—especially since most of these factors will be external to the computer or system.

The question arises, "If these factors are beyond the control of the manufacturers, then what can they do about them?" As noted earlier, these factors can have a direct influence on the reliability of the computer and may not be "excluded as warranty-able failures" for warranty purposes. But interestingly enough, in many cases the manufacturer (or rather, the designer) *can* do something about them.

"Controlled Conditions" versus Murphy's Law

My philosophy here is: Murphy was an optimist—Murphy's Law is conservative. What *can* happen is *certain to*. Anything that is unforeseen *will* take place.

Most people make a joke out of the so-called Murphy's Law, and it *is* funny . . . on the surface. But I have looked beyond the "funny" and have realized that for the most part, it holds a very definite message. It is, in a way, a folktale. But Murphy's Law is conservative. For instance, in the two misquoted laws above, I read the following: "What can happen will." This famous Murphy's Law says to me that anything left to itself will take a turn for the worst. This is so true, it's scary. Think of it this way. Any man-made site on earth, left to itself, will degrade or decay to a "natural" state.

For example, have you ever lived by an old farm with a barn that was left to itself . . . with no repair, no maintenance, and no use? The barn gradually weathers, boards fall off, the roof sags, gradually the barn leans to one side, and eventually it collapses into a pile of weed-grown rubble. After a few more years, the wood rots away, and all that is left is a small mound of dirt where the barn used to be.

This philosophy or principle can be applied to our subject at hand, and indeed becomes very real under the light of close scrutiny. If you leave anything to chance, the swing will be toward what you least expect or want to happen. What *can* happen *will.*

And I'm speaking from experience. Believe me, if you design for optimal conditions, you're in for a very nasty surprise. Don't do it. Design for the worst possibilities, and you will never have to regret overlooking them.

Case in Point

There is an old saying about pilots that is applicable here, in a way: "There are old pilots, and there are bold pilots. But there are no old, bold pilots."

What this implies is that you can get away with one mistake. . . possibly two. But make three and you're dead. Because, for the pilot there are three very important factors: (1) weather, (2) knowing where you are at all times, and (3) maintaining plenty of fuel supply.

You can get away with making one mistake—overlook one of these factors . . . maybe even two. But make the mistake of combining all three factors, and you are dead.

For instance, you might be flying along and (1) run into bad weather. As long as you (2) know where you are, and (3) have plenty of fuel aboard, you can turn around and fly back out of the bad weather. If you run into bad weather and are lost, you are in trouble. But add to these the fact that you are low on fuel, and you are dead.

Or, our second example: say you run low on fuel. As long as you are in good weather and know where you are, you can plot a course to a safe landing where you can obtain fuel. But add to being low on fuel the fact that you don't know where you are, and you're in trouble. Add the fact that you are now in bad weather, and you're dead.

For the third example, assume you are lost. If the weather holds out, and you have plenty of fuel, you have several courses of action that will get you out of trouble, even though you don't know exactly where you are. But add to that being low on fuel, and now you're in trouble. If you now get into bad weather besides, you're dead.

Just remember—as sure as the sun comes up in the morning, the factor you fudge on will be the one that bites you. Design for the worst, hope for the best, and you will be all set if everything falls in between. Look very carefully at all the environmental factors we talk about here. Consider them and their consequences carefully.

The factors we talk about here are things that, as we mentioned, are not within the control of the design engineer or manufacturer—but can wreak havoc with a design that does not take them into account.

A list of these factors might include such interesting and varied (and unexpected) subjects as

1. Weather
    a. Thunderstorm activity
    b. Humidity
    c. Temperature "norms"
    d. Other weather-related environmental factors
2. Soil composition
    a. Annual rainfall amounts
    b. Electrical soil conductivity

3. Natural resources
4. AC power
5. Facility environment

"What could factors such as soil composition or natural resources possibly have to do with the reliability of computer systems?" you ask. But affect reliability they will. Let's start by looking at each item on the list and consider each of them in order.

Factor 1—Weather

You might wonder what effects weather could have on a computer system that is safe and sound inside a cool, dry facility, right? (And if the interior computer environmental requirements are strictly specified and the specs are adhered to, it may not.) Then again . . .

After reading the following, consider very carefully what you have read. Hopefully you should begin to see that indeed, the listed factors below *could* affect the computer . . . even though it *is* inside a "controlled cool, dry facility environment." Because believe me, they do.

A list of some of the factors we speak of here might include

a. Thunderstorm activity.

We have discussed this item before, remember? Thunderstorms themselves probably need no further explanation, but we do have some other comments to add at this point. It is not so much a question of *how much* thunderstorm activity there is in the general vicinity of a computer installation, but what *effects* it may cause or bring with it. For instance, a thunderstorm might cause some of these environmental effects:

1. Changes in humidity—If not specifically and rigidly controlled, the humidity inside the computer room can be directly affected by the outside environment during passage of weather frontal activity, seasonal changes, dry spells, etc. It has already been shown that if humidity is not closely controlled, conditions conducive to electrostatic discharge can occur. By the same token, too high a

humidity contributes to corrosion or even the growth of fungus and mildew.
2. Air pressure in the computer room can vary directly with the barometric pressure outside, as a result of weather frontal activity. Air pressure affects such things as the voltage required for an ESD discharge (the distance a spark will jump), and other possibilities.
3. The conductivity of the ground—even though the composition of the soil may remain the same, the ground's conductivity will vary with rainfall, temperature changes, freezing, snowfall, thunderstorm-associated lightning activity, etc. These things result in changes to the ground's effectivity as a ground plane. This, in turn, affects input AC voltage, radiated RFI susceptibility, etc.

Another factor under the heading of weather that is capable of affecting a computer even though it is inside a computer room might be

b. Average relative humidity.

Why should the average relative humidity of a certain location be a cause of computer problems? Think about this: it is possible because the relative humidity directly affects the conductivity of the soil in the immediate vicinity of the building and the computer room contained within it, by being the determining factor about how quickly it dries out. Naturally, dry weather with very little rainfall will cause high ground resistances unless compensated for in some manner. By the same token, high average humidity combined with the fact that it rains very often in some particular place can guarantee good conductivity of a high-acidity soil at that location. On the other hand, high average rainfall can also "leach" the conductivity out of sandy soil.

c. Seasonal temperature "norms"

High seasonal temperatures during the summer months accompanied by very low temperatures during the winter months can directly affect the computer, particularly if the summer months are especially dry . . . due to the effect it has upon the soil's conductivity (which affects the ground's surge impedance).

    d. Other environmental factors such as vibration, air flow, etc.

What could vibration or airflow have to do with computer failure? Actually, vibration can be a very bothersome external source of system problems. For instance, a building very close to an active railway right-of-way can especially be a problem because of vibration from passing trains shaking the ground similar to the effect of an earthquake. This vibration can cause problems with connections that shake loose, cards that become loose in their card cages, etc.

Airflow is important because of the effect it has upon cooling the computer. In other words, even though the temperature may be maintained at some specified level, without proper airflow the computer will not get the cool air in the needed amounts to the right places.

Factor 2—Soil Composition

What could soil composition have to do with this subject? This becomes a little clearer when you consider the following:

    a. Annual rainfall amounts—here again, rainfall directly affects the soil's inherent *electrical properties*, such as,
    b. The earth's electrical conductivity.

Annual rainfall amounts naturally determine the relative "wetness" of the soil, which has a direct bearing on the soil conductivity—or conversely, its surge impedance. Poor soil conductivity gives rise to possible differences in potential in the grounds, which may be severe enough to affect communications between the computer and its peripherals many feet away.

Factor 3—Natural Resources

Lots of iron ore or copper in the soil means that soil will probably be a better conductor than if the soil were composed of mainly decomposed granite (as much of the Western coast's soil is). Forest land will be different in conductivity than wet sand such as is found in Florida, especially if it contains a considerable amount of acid from leaf mould, decayed plants, etc. This is because the acid soil is a better conductor than sandy soil.

Factor 4—AC Power

This subject has been thoroughly covered in foregoing discussions, but needs to be repeated here—because AC power or grounding can be corrupted from *outside* the computer room as well as from within. Or, the AC power source may be other than the normal facility power from your local power company (such as a motor-driven generator, etc.). More on this after we have considered the last point on our list.

Factor 5—Facility Environment

This final factor is more or less a composite of all the discussed contributors to computer system problems that might be considered as external to the computer (but still are not "warranty-excluded failure" sources).

"Okay," you say, "but I still don't see how weather and the soil's composition and humidity and all the rest can be a factor in the *reliability* of a computer system!" (And it may not be inherently obvious—how these factors may contribute to loss of system reliability . . . especially if, again, the interior computer environment is strictly specified and the specs are adhered to.) But these computer-room environmental factors are often overlooked by the installer or site prep people and they definitely *do* affect computer reliability.

So now let's bring it all together, with a complete explanation of how these factors contribute to computer problems— consider carefully the last three of the above factors—natural resources, AC power, and facility environment—and how they affect the computer.

Probably the best way to illustrate the affect these factors can have on a computer is to relate a case in point.

Case In Point 2

To continue in the vein started on AC power earlier, the following brought it home to me. I was called into a customer's place of business in Tampa, Florida to cure a persistent and incessant computer reliability problem.

The user was furious. He had spent some three thousand dollars to make his system more reliable—on the word of one of the company's salesmen—and was still bothered by constant "lock-ups."

After arriving there, I immediately went about fixing all the immediately obvious things (e.g., unplugging electric fans, printing calculators, remote telephone transmitters and radio transmitters, from the same AC power line as the peripherals drew power from; and in fact, removing them from the area). My methodology includes cleaning up all the little obvious things that can obscure the *real problems*—first. After having done this, I could see there was still another factor affecting the system computer. And it seemed electrical power-oriented. But what was the source of the problem?

We set up a Dranetz™ AC voltage monitor on his computer power for 24 hours. It became clear by studying the printout from the monitor that the AC power was not stable in voltage over a particular period of time during the day. After talking to a representative from the power company who claimed they were not responsible (of course), I considered the following facts very carefully. These facts became vital clues to the cause of the problem:

1. The source of the building's AC power was a transformer switchyard situated some 800–1000 ft from the building.
2. There was considerable lightning activity in the vicinity throughout much of the year.
3. The Florida soil was composed mainly of wet sand, which was leached by considerable rainfall.
4. Outside, the humidity was about 85%, and the temperature was in the 80s.

These factors, when each was considered by itself, did not add up to the total problem we were experiencing. But when added together, they spelled T-R-O-U-B-L-E! Here's why that was true, and what was really happening:

Items 1 and 3 above together form a very unstable ground-return path. Why? Consider this: Wet sand is not a very good conductor, especially when all the salt has been leached out by a recent rainfall. A transformer farm in a switchyard as close (or as far away[1]) as this one was is a real problem as far as ground voltage gradients are concerned.

If the surrounding area's AC load varies with time of day (as it would if huge numbers of air conditioners were turned on or off), the voltage gradient in the ground is going to vary as a function of the load (or put another way, it will vary with the amount of power consumed).

To top it off, if there were a thunderstorm anywhere nearby (see item 3, above) it would contribute voltage variations of its own by affecting the AC power-line voltage into the transformer switch-yard itself. This can be a big problem anytime, but is especially so when the local ground soil is a poor conductor . . . particularly at the distances we are talking about here. And it became very obvious when we put a recording voltage monitor on the building's AC entry circuit from the main AC line—the voltage was all over the map, and was dependent upon what time of day it was. I discovered after some poking around that there had been no dedicated circuits or isolated grounds pulled from the power entry into the building for this computer installation.

This is a "no-no." Separate isolated ground wires should be pulled for every computer circuit, and each circuit should have a separate, clean dedicated AC line all the way from the power entry.

Several things had to be done in parallel to solve this problem. First off, the power company made their contribution by making the input power as "stiff" as possible. (We won't go deeply into this, because it is outside the scope of our subject—just bear in mind that you should be sure that the power company maintains the AC voltage within a very close toler-

---

1. Depending on how you look at it, distance from the building can be either, and 800 ft is too close.

ance at the AC power entry into the building your product resides in.)

Second, a power isolation transformer was required to allow a reference ground to be established right at the power entry, from which a *dedicated circuit* was pulled for the computer and its peripherals. Next, a *dedicated, isolated (third wire) ground* was pulled all the way from the AC entry point to the computer *and* the farthest point away that any peripheral existed. A voltage regulator or line conditioner was installed at any peripheral that was more than a few score feet from the computer. The reason for this, of course, was to maintain the voltage gradient between computer and peripheral at as low a voltage differential as possible.

However, here I must point out a rule of thumb: that *two voltage regulators or line conditioners of the same type should never be installed at more than one point on the same circuit!* The reason for this rule is that if two tap-switcher type voltage regulator devices[2] are installed on the same circuit, they "fight" each other by causing either the line voltage or the ground gradient to vary in steps, and if each tap-switcher is a slightly different distance from the source, each of them will switch taps (higher or lower) at different points, thereby causing different input voltages at the two locations.

By the same token, if both devices are ferro-resonant regulators, the load on the line will vary with the voltage that appears at that particular device (which is a function of how far from the source it is and the amount of current required at that particular voltage)—remember, $P = I \times E$, so for P to remain the same, if voltage (E) changes in one direction, current (I) will change in the other direction. Thus, the current load will vary with voltage, which is being manipulated at two different points on the same circuit. Feedback does the rest, and you have an oscillating voltage condition. The frequency of that oscillation is very low, and causes a regular, and even predictable[3] fluctuation in the line voltage at both points that will drive a computer system nuts.

---

2. This is explained in a later chapter.
3. Predictable, as long as the line's *load* does not change in the meantime.

So, What?

How does all this affect the designer of the computer system that resides in a place similar to the one just described? Stay tuned . . .

Now, I will be the first to admit that a good design engineer (no matter how intelligent and experienced he or she is) would be hard pressed to be able to foresee dumb mistakes such as the user or an inexperienced field service engineer using two regulators of the same type on the same branch of an AC circuit and causing an oscillation in the voltage level, or to foresee that the earth would consist of wet sand at the user's location or that the transformer switchyard would be located such that ground gradients could cause trouble as a result.

Requirements Specifications

What I am advocating here is that the design engineer consider all the negative factors that *can* be a part of the final installation, and try to tailor your design to allow the least amount of trouble to result from these factors. Those factors that are beyond your ability to design out (for reasons of cost or other trade-off considerations) must then be set down as *minimum requirements specifications for installation,* to be seen to during site preparation efforts.

This is, of course why it is absolutely necessary for the manufacturer to have a true and complete set of minimum requirements set down in a site preparation requirements specification. The question of whose job it is to write this specification document is, of course, at the discretion of the manufacturer. But if I were asked, it's my opinion that it should be the design engineer who draws up the minimum requirements specifications for the portion of the product he personally designed—even though he might not write the overall specification document itself.

Beyond that, there should be a strict set of warranty-related rules in the user's manual or documentation spelling out the "no-no's" to be observed by the user. And the user would be at risk of voiding the warranty if the rules are violated. Again, these rules should be drawn up by the design engineer for each part of the system (who after all, is the most knowledgeable about all the small details of that portion of the system's design—such as the power supply's requirements, the computer's own special requirements, or the connection, care, and feeding of peripherals). These individual requirements should then be compiled into a system spec.

The reason some manufacturers do not set up rigid specifications is that the sales department doesn't want anything said about the system that could be considered a negative. This is silly, in my opinion, because to sell a system and not tell the user all the facts up front will guarantee a disgusted user when they do eventually find out all the facts.

Once a rigid set of site preparation requirements have been drawn up and established as minimum installation requirements, they should be published as such and given to anyone responsible for installing systems. Design engineers can then breathe easier in the knowledge that the factors beyond their control are now at least specified as minimum requirements. They can then turn their attention to reliability factors that *are within* their control.

Installation Site Considerations

I very recently ran across a situation that I believe is worthy of further discussion here. It represents the challenge one sometimes runs across in adapting a computer to its intended environment.

During the process of putting this book together, my wife and I bought a 34-ft motorhome. It quickly became obvious to me that if I were to get any writing done, I needed to take my Macintosh computer along on the trips in the motorhome that we made at every opportunity (such as on weekends and vacations). I wanted to relax and write while we were away from home, in the relaxing atmosphere of our motorhome in an RV park somewhere.

Now, a motorhome is not exactly your ideal computer location site. But if I were to compile a list of hostile environments for small computers, it would probably look something like this (best first):

1. Truck, van, or motorhome
    - Very unreliable noisy and surge-susceptible AC power source
    - Electrically noisy environment (engine ignition, DC alternator, AC generator, etc.)
    - Lack of good solid ground plane
    - Electrically noisy appliances
2. Small boats
    - All the above-mentioned, plus
    - Salt air
    - Bad AC power
    - Electrically noisy environment with highly magnetic equipment close by
3. Welding shop
    - Huge RFI and EMI sources
    - Contamination from smoke and carbon products
4. Power station or switchyard
    - Large EMI sources
    - Electrically noisy equipment
    - Huge sources of RFI
    - Large ground gradients
5. Auto repair shop
    - Electrically noisy from all sources above
    - Huge sources of EMI from grinders, wheel balancers, drills, air compressors, etc.
6. Machine shop
    - Electrically noisy equipment
    - Huge EMI sources
    - ESD problems as well
7. Industrial manufacturing plant

Of course, this list is arbitrary, due to a large number of variables involved. And number one and two are a close race. But you get the idea.

As I said, immediately after making our motorhome purchase, I started investigating the motorhome as an environment for a desktop computer. The more I looked at the problem, the more it became obvious that a motorhome is one of the worst environments imaginable for a small computer. The AC power available from the AC generator[4] is very "soft," indeed.[5] The AC power available even when the coach is connected to external facility power (at an RV park, etc.) is not by any means clean and steady.

At any rate, it required a considerable amount of preparation to make the coach acceptable as a site for the Macintosh. For the sake of example, let me take you through the site-prep requirements for installing a computer (or computer-driven appliance) in a motorhome.

The Motorhome as a Computer Environment

A motorhome electrical system is usually set up with from two to five 12-V storage batteries. In a self-contained motorhome, one of these will be a high-current and high-ampere-hour-capacity battery for use with the engine starter, which is recharged through a "battery isolator" from the engine-driven DC-output "alternator."[6] One side of the battery isolator charges the "engine battery." The other side of the isolator is connected to, and charges the "house batteries" (see next paragraph). The center input of the battery isolator is connected to the engine's alternator.

12 V DC from the "house batteries" is routed to all the 12-V accessories on the motorhome, including all running lights, interior lights, refrigerator (if capable of 12-V operation), water pump, and the automatic drop-step (if present), etc. Also con-

---

4. The self-contained AC power supply is called the "AC generator" by RV enthusiasts. But "generator" in this context is a misnomer. A generator is usually used to supply DC.

5. A "soft" power supply is one whose power output is limited in capacity, causing voltage to go down as a function of the amount of current required.

6. "Alternator" used in this context is also a misnomer. By definition, an "Alternator" puts out *alternating voltage*, while the engine-driven "generator" puts out DC.

nected to this bus are the ignition circuits for the hot-water heater, the furnace and its blower (if there is one), and so on.

There is usually an AC power "generator" that is driven by a small gas- or propane-powered engine. This AC source supplies AC power to all 120-V AC accessories such as the TV set, microwave oven (if supplied), the air conditioners on the roof, the AC-to-DC converter for charging the "house" batteries, and possibly the refrigerator, etc.

Now, the fact is that the coach-contained AC "generator" is not an infinite source of current (they are usually rated from 2500 to 7500 W at 120 V AC, and are therefore limited in current output—a very "soft" source, indeed). And since they are *frequency*-regulated mechanically—rather than *voltage*-regulated, the output voltage may vary all over the map . . . from 109 to 135 V RMS, depending on the load. But the AC frequency also tends to vary somewhat, which is caused mainly by the amount of mechanical delay in the regulating circuit. The result of all this is that due to sags, surges, switching transients, and other electrical noise, this AC is not a good source of power for a computer.

Therefore the question arises, should a motorhome owner rely on the AC generator to supply varying-voltage AC power—possibly to be conditioned by a power conditioner before consumption by the computer? Or should another alternative be found? Most computers do not like varying AC input line voltage at all (not to mention that this also means that RV owners must run the AC generator at the cost of some one gallon per hour of gasoline consumption; ideally, they should be able to draw AC power anytime they need it). Considering all these things, I would say the answer to the question above is "no." But what are the alternatives? Is there another way? Yes, there are really two other ways.

Uninterruptible Power—An Alternative

We are going into this subject here as though it is related to motorhomes, but the following can and should also be applied to any other application with the same type of problem.

There are times when AC power from the coach's AC generator fails (or the fuel supply runs low and the user decides to turn it off)—and the AC power is off for more than just a few seconds. If the computer is in use at that time, this kind of outage can be disastrous. This problem probably cannot at present be inexpensively solved by the design engineer at the board level. Yet the problem does exist, and can sometimes be the difference between customer satisfaction with the product and continuing problems that lead to total rejection of the product by the user.

About the only known economical cure for this problem at this time would be what's known in the trade as an *uninterruptible power supply* (UPS). What is a UPS, and what does it do?

A UPS is a battery-powered AC inverter, combined with a method of keeping its battery charged. Besides the input power source and the AC outlet, every UPS consists of at least three basic modules, as shown in Figure 6.1: (1) the AC–DC converter/battery-charger module, (2) the DC storage batteries that supply power to (3) the third module, the DC-to-AC inverter module.

Figure 6.1 shows these elements or modules and their relationship to each other. In this diagram, all modules are directly connected to each other. In reality, there are different ways to connect these modules, as shown later.

In the example shown in Figure 6.1, AC power from the AC generator is converted to DC in the first module (called a "con-

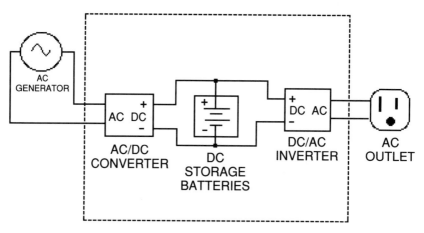

Figure 6.1 UPS elements.

verter"), then this DC is used to charge the batteries in the second module. The batteries are usually 24 or 48 V, and they merely supply current to the third module . . . the DC-to-AC "inverter."[7]

Actually, a UPS can be one of two types—which differ mainly in the way in which they supply AC to the load. As mentioned in an earlier chapter, the two types of UPS are the *on-line* type and the *switch-over* type. This difference between the two may be ever so subtle, but it is very important.

An on-line UPS consists of the same three basic UPS elements or "modules" as those shown in Figure 6.1, but differs in how it is connected. In this type, the input AC power never actually goes to the load. Instead, input AC power is used simply for recharging the parts that make up the second module—the batteries. The third module is the inverter, which actually does the job of supplying 60 Hz AC to the load, drawing its power input DC from the batteries. The on-line UPS scheme is diagrammed in Figure 6.2.

In an on-line UPS, the output DC-to-AC inverter *always* draws power from the batteries, even when AC input power is available. The batteries are recharged from available input AC through the AC-to-DC converter only during the time when AC

7. An inverter converts a low (usually 12–48 V) DC voltage to 120 V AC.

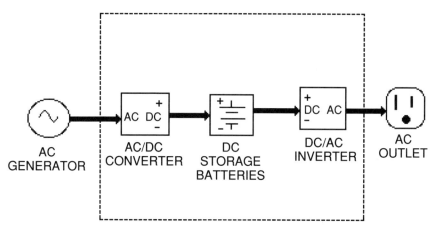

Figure 6.2 The on-line UPS scheme.

input is available. When power input fails, the batteries simply continue to deliver current to the inverter but *are not being recharged*, and will eventually "run down" if power does not come back on within the limits of the ampere-hour capacity of the batteries. The AC supply is never connected directly to the load, so there is no "switch-over" required.

On the other hand, the second type or switch-over UPS consists of the same basic three modules as the on-line type diagrammed in Figure 6.2. But the switch-over UPS differs from the on-line type by the addition of a fourth module—a transfer switch circuit and a slightly different wiring scheme. The switch-over-type UPS, with its fourth module and the switch-over or transfer switch circuitry, is shown in diagram form in Figure 6.3.

The switch-over UPS utilizes the AC power direct from the power company to supply the load, until that power source becomes unreliable or fails. At that moment, the UPS *switches modes* from the AC power line supplying the load, to an internal battery-supported inverter supplying power to the load. The switch-over UPS is usually a little cheaper than the on-line type. . . not because it contains fewer parts, necessarily, but more because it is not as fool-proof and "invisible" at switch-over time. By that I mean, when a power failure occurs, the switch-over circuit actually takes a small amount of time to do

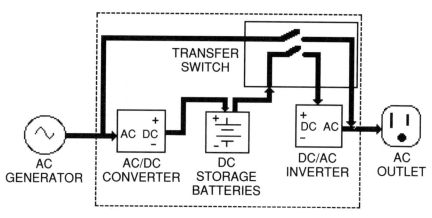

Figure 6.3 The switch-over UPS.

the switch from facility power to the internal battery-powered inverter. This small amount of time may be very short, but still can cause such a glitch as to make a very sensitive computer "hiccup," or even die. This will probably happen even if the output sine wave is synchronized to the input sine wave at the time of the switch-over (see Fig. 6.4).

What is even worse, the switch-over can occur at any time—from the zero-crossover point to the highest maximum voltage excursion of either alternation. The diagram in Figure 6.4 shows the switch-over occurring soon after the maximum positive voltage excursion of the AC sine wave.

An even more important consideration is *the output waveshape* of the inverter-module-supplied AC.

Effects of Inverter Output Waveshape

Nearly all the less expensive types of UPS supply an alternating *square-wave output.* That is, the output square wave is alternating in polarity but is a square pulse rather than a true AC sine wave. This is not a "clean" computer-grade power to be used for either the computer or its peripherals, because of the huge amount of high-frequency components inherent in the rapid rise and fall times of the square wave. Recall here that the high frequencies affect computers the most. Therefore, it is unlikely that a square-wave-output inverter will serve the purpose without problems.

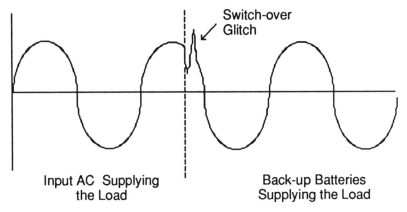

Figure 6.4 The "switch-over glitch."

Be aware, however, that *square-wave outputs are commonly in use* . . . in both the on-line and the switch-over-variety UPS devices. My advice is, be certain that the output of the UPS selected is truly a "sine-wave" output. It may cost a little more, but it is definitely worth it. Do not settle for a square-wave output if you are truly concerned about your computer's welfare.

An Alternative?

There is one other way . . . a second but seemingly better option. The simple addition of one section: an *inverter* . . . the only *other* (but very expensive) way of getting 120 V AC from available DC power. Aside from a UPS, the inverter is the best way to have AC power available without running a noisy AC generator. For this reason, some of the new motorhomes are coming out with an inverter *instead of* a gasoline-driven AC generator.

Now, it happens that two of the modules contained in a UPS are *already present* in a motorhome. These are (besides the AC generator): the AC-to-DC converter and the storage batteries—in this case, the "house" batteries. The AC generator or alternator recharges these batteries anytime input power is available. Of course, the engine alternator also recharges these batteries whenever it is running.

Figure 6.5 shows the elements of the UPS already available

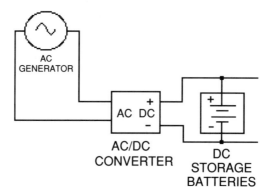

Figure 6.5 Available elements.

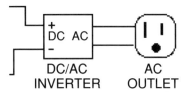

Figure 6.6 Additional elements required.

as installed equipment in the motorhome: the AC power source, the AC-to-DC converter, and the DC storage batteries.

Since the motorhome contains two-thirds of a complete UPS[8] anyway (see Fig. 6.2), why shouldn't the owner merely add the third section required, in the form of an inverter as shown in Figure 6.6? This appears to be a simple addition but turns out to be a very expensive one in terms of parts cost. It seems these sine-wave inverters are not as cheap to manufacture as a square-wave type, due to the high-power AC-handling components required.

Because of the high initial cost, at this time only a very few manufacturers supply motorhomes equipped with anything other than AC-generator plants as the source of AC power. Furthermore, as pointed out, few electronics manufacturers supply large-capacity inverters that have a sine-wave output (especially with a 12-V DC input). On the other hand, a great many manufacturers supply 12-V-DC-to-120-V-AC inverters whose output is a *square wave*. A computer's switching power supply being fed from a square-wave output may cause many problems.

The voltage from a square-wave inverter may be alternating in direction, but being a square wave, it remains at the high point in the cycle in both directions for too long at a time (which may cause very high current consumption and overheating of the switching power supply, thereby greatly reducing its active life span).

If a square-wave output inverter is used to supply AC to a computer's switching power supply, its output should by all means be "smoothed out" by some sort of a choke–capacitor

---

8. A complete UPS is shown in Figure 6.1, showing the three components in a UPS.

filter, and/or be put through an EMI filter prior to the switching power supply before being fed to the computer. This may not be a practical, viable solution to the problem, for many reasons.

Conclusions

The conclusions this author reached after extensive testing, are these final facts . . . backed up by actual research:

1. The computer should be driven from an *inverter* that feeds continually from the 12-V-DC system in the motorhome. But the output voltage should be a sine wave from the inverter. In this way, the coach owner is not required to run the AC "generator" for hours on end while using the computer.
2. Grounding of the computer's power supply is extremely important. A separate, isolated ground wire from the inverter directly to the vehicle's frame is extremely important here.
3. EMI and RFI filtering should be a part of the AC power's route into the computer. This can best be done with a line filter or line conditioner. The filter described in Chapter 3 is ideal for this application.
4. The computer itself, plus any peripherals must be connected to this clean, dedicated AC power supply. This is important to prevent ground loops and noise sources, as described in this book. Therefore, when installing a computer into an environment such as a motorhome, observe these rules of thumb, and you should be all set.

Happy camping!

The Boat as a Computer Facility

The boat presents a similar, but in some respects a more demanding site for a computer installation. If used on a boat, not only is there the problem of creating clean power but there is also the consideration of salt-air corrosion, high humidity and/or actual wetness, high incidental magnetic interference,

RFI from such sources as the shipboard radio and electrical storms, etc.

The one saving grace about a boat is that there is usually an opportunity for a good ground plane. In fact, if done properly, the water itself can be part of the ground plane. A metal plate under the water line is usually used, connected with separate wires to every piece of metal on the boat. This metal plate then becomes the "single-point ground" for the entire boat.

The Welding Shop as a Computer Site

A welding shop is another very demanding computer site. The welding environment is rich in

- Huge RFI and EMI sources
- Contamination from smoke and carbon products

The EMI and RFI from the arc-welder or plasma torch requires adequate shielding for the computer itself (especially the disk drives), and the I/O cables and all peripherals. Special care must be taken in all these areas.

Grounding is of utmost importance. Single-point (or as they are called—"equipotential grounds" must be established.

The Power Station or Switchyard

A power station or switchyard is yet another bothersome environment in which to install a computer, unless special precautions are taken to overcome the effects of

- Large EMI sources
- Electrically noisy equipment
- Huge sources of RFI
- Large ground gradients.

These effects result from the natural switching transients and magnetic fields generated at a power switching station or transformer yard. A substantial amount of shielding is required, as well as a dedicated, isolated power source for the computer itself.

The Auto Repair Shop

As was detailed in an earlier chapter, an automobile repair shop harbors all sorts of "monsters" that like to eat computers for lunch. Among these are huge sources of EMI such as grinders, wheel balancers, drills, and air compressors.

The Machine Shop

Included in a machine shop are also many of the same sources of EMI and RFI, electrically noisy equipment that bring about EMI sources. ESD problems abound, as well, since most repair shops are part of a large retail new-car dealership, which are heavily carpeted.

The Industrial Manufacturing Plant

Without reiterating all the points mentioned above, suffice it to say here that since industrial manufacturing plants have large machinery, they have all the foregoing listed problems, as well as the accompanying EMI problems, plus surge and sag, as well.

Remember the principle of I/O problems from the last chapter—I/O problems are often just "ignorant operator" problems. Do your homework ahead of time. Be alert to the possibilities your product may be subjected to. Make your product pass the test. Design against EMI, RFI, and ESD up front. And above all, write a specification to maintain environmental conditions to a manageable level, and make the warranty dependent on the specified conditions existing.

Please understand—I am not trying to give you hard-and-fast, cut-and-dried rules to follow in design and implementation. I would not insult your intelligence by suggesting that I can give you a hard-and-fast set of rules to follow that will be a cure-all. What I am attempting to do is acquaint you with a philosophy—to give you a background of information and the resulting insight it takes to foresee possible outcomes of not designing against the possibilities . . . the inevitable. Yes, I believe that the larger the numbers of a given device manufactured, the greater the chance that every negative possibility will eventually happen. Whether or not it is worth the effort to design these possibilities out is ultimately up to you.

If you are not a designer or an engineer, at least reading this may give you an idea of what they face when trying to design a given piece of equipment . . . the ultimate feeling of futility they must experience when a design goes wrong because they were human and failed to foresee the inevitable.

CHAPTER 7

Why "Burn-in"?

If you've never been part of a component-level or board-level production environment, you're probably asking yourself, "What has *burn-in*[1] got to do with the subject of ESD, RFI, and EMI?" And right you are to ask this question, of course! But this book would not be complete without a discussion of "burn-in," since it has such a direct bearing on system reliability—at least in the early life of a system.

For those readers who are wondering "what *is* burn-in, anyway?" we will for the purposes of this book define burn-in as the process of "cooking" new parts at elevated operating temperatures, with normal or increased operating voltages applied for the length of time required to assure ourselves that they will continue to work without premature failure due to "infant mortality."[2] Quite a sentence, eh? Seriously, the burning-in of parts has become a very scientific part of our field of endeavor. A lot of developmental research has been done on just the equipment and technologies involved.

It turns out, as a result of all this scientific investigation, that there is a period in every young integrated circuit's (IC's) life in which it is subject to the "crib death syndrome" or *infant*

---

1. "Burn-in" is the period of elevated temperature operation for the purpose of eliminating early failures (see Glossary).
2. Infant mortality is defined herein as failure of a "new" part during the first 72 hours of its operation.

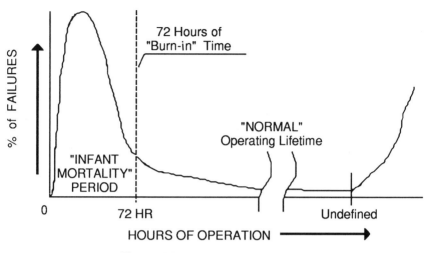

Figure 7.1 "Infant mortality" period.

mortality. This involves the probability of failure, during the period that starts when an IC is first installed and power is applied, until some number of hours later. The most critical time in an IC's life is probably during the first 72 hours.[3] Until at least this period has expired, there is more than a slight chance that it will fail.

Said another way, if a *new* IC is going to fail under operating conditions, the chances are it will fail during the first 72 hours. Now, some IC manufacturers or their employees may differ with regard to the length of this period and/or the exact high point in the curve. In fact, you would probably get a different answer from any person you ask, since each may have a different set of statistics they go by, or the particular IC in question may not exhibit this phenomenon. But we are talking here of *practical* original equipment manufacturer (OEM) or end-user experience.

Figure 7.1 shows a graph of what the lifetimes of a typical IC type might look like. You will notice that the maximum number of "deaths" peaks somewhere around the very first few hours of operation. Lots of ICs never get this far, of

---

3. At least, in the case of TTL (transistor–transistor logic) parts we have had experience with.

course, and are "dead-on-arrival" as soon as they are manufactured or received. Here, we are not counting those—only the ICs that get far enough to be inserted in a circuit and power applied. As shown in Figure 7.1, we experience the highest percentage of failures somewhere within the period of from 0 to 72 hours of operation.

It has been proved that semiconductor products subjected to "burn-in" periods of a sufficient length of time show fewer failures after delivery to customers than those not burned in. Furthermore, it has also been proven that the later in the OEM assembly process a part fails, the more expensive it is to fix, and it is many times cheaper early in the assembly process to eliminate parts that are destined to fail.

The period of time allotted for the burn-in process[4] varies from 12 to 168 hours, from manufacturer to manufacturer (also dependent on whether the manufacturer is the original product manufacturer or the OEM user of the product). It stands to reason that some kind of sorting test to eliminate parts that have failed should be run during the burn-in, but many of these manufacturers do not test the part until it has completed its burn-in cycle. If the parts are powered up all during the burn-in and are continually tested, they can be considered "seasoned parts" and have arrived at the lower portion of the failure curve by the time the period has expired. Having been monitored and fully tested during the burn-in cycle, they can then be used in a next assembly (usually a printed-circuit board of some sort) with some assurance that they will avert the more expensive possibility of failure as a soldered part in a printed-circuit board—or even worse, after the board has been installed and checked out as a part of a system.

Called TDBI (test during burn-in) by some, this procedure is easily seen as much more efficient than burning in such assemblies as complete printed circuit boards . . . which must have failed parts desoldered and resoldered if they fail later. In this method, the parts are installed in a test fixture, which applies

---

4. This length of time is dependent on the exact product being "burned in." For example, boards that contain ICs that have not been "burned in" will require more time in the board-level burn-in process than a board with "seasoned" ICs installed.

the proper inputs and actually checks the output pins for the correct signal outputs. Any and all failure data can easily be saved as a database and yields much in the way of very important information about failure modes, trends, process problems, various manufactures versus part failures of their particular product, etc. The length of time for burn-in can be adjusted to an optimum for each particular type of part using accumulated data.

It is true, however, that this method requires very sophisticated automated test equipment (ATE) to run dynamic tests at the parts' specified speeds while burn-in is in progress. And the fact is that the fixtures used for burn-in require extended cables or wiring to get signals in and out. These factors increase the cost of this type of early product-failure detection. But the results of early detection of uncertain parts far outweigh these small increases in cost—especially when it is considered that failures detected by this means eliminate the *thermal-intermittent*[5] parts (those parts that tend to fail under stress but return to operable condition when the stress is removed). In fact, environmental or burn-in chamber temperatures should actually be cycled up and down in temperature during the part's burn-in period, in view of the fact that thermal shock or stress is an important cause of early failures in components. Testing should be maintained during the thermal cycling to guarantee that thermal-intermittent problems are discovered that would go undetected if they "healed" by themselves after the thermal stress was removed. Early detection of thermal faults or failures is worth its weight in gold, when compared to the hours of laborious testing required to find a thermal problem in a printed-circuit board. This category alone could save the company the added cost of the testing.

Another added advantage of the early temperature cycling and dynamic testing of parts is that extended diagnostic testing of boards that failed (due to "infant mortality") is no longer the problem that it was without these tests. It is not at all unusual for a technician to spend several hours on a single board to find

---

5. Thermal-intermittents are those problems that show up under elevated or reduced temperatures and "go away" when the part is returned to normal temperature.

a thermal-intermittent problem. This is a very expensive way to find and replace a bad part.

The "Skimming of Good Parts" Scenario

The scenario *without* dynamic testing during temperature cycling usually goes like this: Let's say that the production area is in a hurry for a certain type of product . . . possibly a printed circuit board of a specific type. So the test technicians "skim off" the best of the product yield, in order to satisfy the immediate shipping requirements. Soon, however, they come to the end of the "cream of the crop" and start looking at those parts that barely failed to meet minimum specifications. If the product is at a printed-circuit-board level, the technician will then sort through these to see if it is possible to "make them pass." Then the technician will search (by lightly troubleshooting) through the remainder of "bad" boards to find those that are the easiest to fix, and if the problem is not immediately obvious will set that board aside and go on to the next. In this way, the second level of complexity is brought into shippable product.

Then, if the demand for this part continues, the technician will be forced to delve one level deeper, and spend more time per board in finding and fixing the more complex problems. If a board takes too much time in this effort, the technician will set it aside once more, and it becomes what is termed in the business as a "dog"—a board that has a problem so hard to find that it requires more time and effort than the technician can afford to spend on a production item.

I have even seen companies where at this point, the demand for production units was so great that the technicians were forced by higher management to take those boards that did not pass all the tests (but would work well enough to "get by") and put them into a system to "get it shipped."

As can be seen, if boards with early failures are "culled out" during the burn-in cycle, they will not find themselves being incorporated into a product—only to fail further along in the production cycle. It is not hard to understand, then, that it makes it much less expensive to find the problem before it is a part of a more complex product. This is the objective of early testing during burn-in.

This type of problem is usually further modified if ATE is used, since the product either passes or fails during the automated test. This makes the first-pass test much faster and more accurate. But from that point on, there still must be a human troubleshooter that finds the problem, and either fixes it personally or points out the part(s) to be replaced and passes it to a rework area. Once fixed, the part can be returned to the ATE machine for retesting.

There is ATE available for testing a product at all stages of the product's utilization—from the part's original manufacture, all the way to the final product that utilizes the IC (which may be a large system that incorporates the board-level product). Manufacture and development of ATE is a very large market these days, and is rapidly growing.

Beyond ATE, there is growing concern about manufacturing products that can either test themselves, or which incorporate testability as a feature. I recommend that anyone who is concerned with the design of a product, design *testability*[6] into the product. This is necessary for several reasons. For instance, the more testability that is designed into the product, the cheaper it will be to put out a reliable product. And after all—profit is the name of the game.

Testability can be incorporated at several points in a product's life. In earlier days, board-level tests utilizing "bed-of-nails"-type testers required that individual ICs on a board be exercised by inputting a signal and checking a response at the output. The problem was that in many more instances than not, this particular IC was "buried" so deeply in the board's circuitry that it meant pulling down a previous IC's active output (which might at that particular moment be "high") or the reverse—forcing a "low" output "high" to get the correct state out) . . . risky business at best!

If you know anything about the internal construction of ICs,

---

6. Testability is defined here as ease of testing of part during the manufacturing process.

you *know* that this is risky business—because this means that a "low" output (which is near ground potential) must be somehow pulled up into the "high" state . . . either by being saturated with current to the point where the voltage is higher than the threshold, or by some other means. Usually (but not always), current is limited when a "high" output is pulled "low", but this is not necessarily true when a "low" output is forced "high." The result can be a failed part that was overstressed in *test*—which is worse than the part failing due to "infant mortality." Because in causing this particular IC to fail, the test may also have overstressed other parts. And they may not show the resulting weakness at the time but will fail under temperature shock or other induced stress later on in life.

This happens all too often. The scenario related here is responsible for an estimated 50% of early failures in the ultimate product. By the way, this is even further reason for doing elevated-temperature burn-in and for *proper testing* of the board-level product before shipping or utilizing it. By "proper" testing, I mean testing that does not force active outputs to the opposite state, causing them to be overstressed in the process.

As to testability, as I said earlier, it can and should be "built into" the product. Let's take a hypothetical case, and run this out.

Case in Point

Let's say that our product for this first example is an IC. Designing in testability is the most difficult at this stage, for the simple reason that to provide alternate inputs and outputs means additional pins. Additional pins mean more board real-estate requirements, as well as cost. But here, internal buffering can mean the difference between success and failure for an IC.

For example, there was once a multiple "flip-flop"[7] IC (made by a company who shall remain unnamed) that had unbuffered[8] outputs. The result of course, was that any change that

---

7. A flip-flop is a type of IC that accepts up to two inputs and a clock. At the time the clock pulse appears, the state of the inputs determine the state the outputs will assume.

8. Unbuffered means the outputs were tied directly to the flip-flop portion and did not incorporate a separate stage to isolate the output from the flip-flop itself.

might take place at one output of the device (such as a failure in the following input "shorting" the "Q" output to ground), caused an actual change of state *at both its outputs* by actually "flipping" the flip-flop to the other state. This is a very undesirable trait in a part.

To have discovered this failure mode early on would have saved this manufacturer from having his parts barred from the procurement lists of many users. But they were not caught early in many cases, and the resulting failures later in the life of the part earned this part a bad name. It even led to the parts being banned in many products. The company later recovered reliability by placing a revised version of the part on the market and recalling the old ones, but it was a very expensive lesson for everyone concerned. This can be disastrous to a small company.

Built-in Testability

At the board level, testability is much more easily built into the product. For instance, in order to test a complete *defined function* on a board, it is necessary to know only the possible input states that cause defined output states.[9] We then test the board by causing those states to exist . . . while looking for the proper output under every possible condition. Sounds easy, doesn't it? Unfortunately, it isn't quite as easy as it sounds. This is because there are usually many other inputs along the way whose state must also be defined for every possible condition. But it is not impossible, especially if special pains are taken during design to "build-in" testability.

In the above example, it might be necessary to provide test points at which the required inputs can be provided from the tester under prescribed conditions—to force the outputs to assume the required states.

Testability should be designed into circuit boards as well as integrated circuits . . . so that they can be properly and easily tested. Testability should also be designed into a computer at the system level—with provisions to input signals or com-

---

9. Known as a "truth-table."

mands in such a way that they prove out every path that can be exercised.

Actually, testability is a field of endeavor in itself and is probably beyond the scope of this book. Our purpose here is to merely point out that this technology exists, and to urge the design engineer to become familiar with and involved in its use.

So if you do not presently use this approach, please look into it . . . become involved, and learn to build in the maximum testability within budget, real estate, and other constraints forced on the designer. Your products will be the better for it.

Back to the subject at hand, the amount of time required for burn-in. This amount of burn-in time varies widely by component type. Generally burn-in time is longer for a new printed-circuit (PC) board that has just been built ("stuffed and wave-soldered," in the case of through-hole board technology, or "stuck & soldered" in the case of surface mount technology) than for a system that has just been assembled with boards that have already been burned in.

Testing time can be subtracted from the required time to be spent in the burn-in rack, since power is on during this period. But time in the burn-in rack can be combined with dynamic power-on testing, or at least exercising the boards. Thus the time spent will have accomplished both ends. Two for one ain't bad!

Using this technique is really cheaper in the long run (regardless of what Marketing or the "bean-counters" tell you). Because as we said earlier, failures caught early in the manufacturing process are many times cheaper than those found in a system that has been shipped to a customer. If you buy boards from a board manufacturing company to be integrated into a system, be sure you give them the amount of burn-in time (under the extremes in operating conditions) before they are passed and sent into the final assembly area.

Remember this old adage—"You'll save lotsa money if you do, but you'll spend lots more than you would have spent for burn-in by replacing bad boards in the field if you *don't*—besides risking the confidence of your customer!" . . . And as we've seen many times before, a customer, once burned, is very hard to win back. Nobody can improve business by losing customers!

CHAPTER 8

Environmental Effects and ESD

This chapter features an enlightening discussion of some other environmental factors—such as high and low ambient temperatures, humidity, and ESD . . . and the effect they have on performance. In this chapter, we will learn about the effects of high component surface temperatures due to power dissipation and the resulting effects on the requirement for cooling air supply expressed as volume. We will discuss some methods of eliminating, or at least controlling these effects.

Ambient Temperature

Ambient temperatures[1] *inside* a system enclosure when a system is powered-down and "stone cold" or "dead cold" will quite naturally be about the same as the temperature *outside* the enclosure. But when system power is applied and the components reach their final operating temperatures, "ambient" temperature assumes a new meaning, and must be further divided into two separate categories.

---

1. Ambient temperatures here refer to those temperatures found inside the housing or enclosure of a typical system.

Category One—Cooling Air Supply Temperature

The first category is the temperature of the external *cooling air supplied to* (or ingested by) *the enclosure*. This first "ambient" temperature is not necessarily in direct contact with the components of the printed-circuit boards. The reason for this is as follows: The cooling air supply may be 60°F as supplied from the air conditioning into the enclosure, but may be warmed on its way through the enclosure from other boards or sources of heat before passing by the components on another board.

Category Two—Enclosure Air Supply Temperature

The second category is the temperature of the air that is *passing by the components* on a particular printed-circuit board—inside the enclosure. This second category may or may not be (and probably won't be) the same temperature as the *supply* cooling air—i.e., the second category is really not the ambient available (incoming) cooling air temperature in the computer room, but the surface air temperature around the components inside the enclosure under operating conditions.

These two separate types of temperatures (categories) are better illustrated and understood by taking a look at Figure 8.1.

Still another important temperature is that found at the components surface, known as *component surface temperatures* under operating conditions. The surface temperature of an individual component under operating conditions is really the most important temperature. These operating conditions would, of course, include any fans or other air-moving and/or cooling equipment in full operation.

Deep inside the integrated circuit in this example, there are forces at work that cause some of the current being drawn by the circuit to be converted to heat. This heat is dissipated by temperature gradients that exist between the proximity of the circuit chip itself and the air passing by the surface of the IC

Ambient Temperature / 159

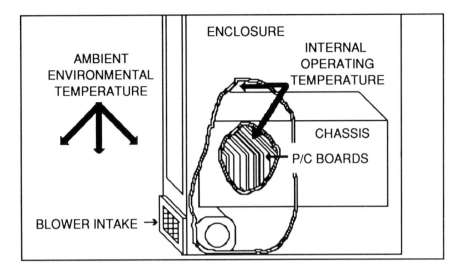

Figure 8.1 Temperatures. These two categories of "ambient" temperature are usually differentiated as follows: the air supply *outside* the enclosure can be specified as the *"ambient environmental temperature,"* while the temperatures found close to the surface of components on the board (and modified by the cooling air moving across the components) are referred to as *internal operating temperatures.*

package.[2] This temperature gradient exists in the compound used for the IC package, and causes flow of heat to the surface of the integrated circuit, where it can be drawn away by the stream of cooling air. Naturally, the temperature gradient varies from IC type to IC type (and, indeed, between two ICs of the same type), due to a multitude of variables[3].

Part of the system designer's job is to think about the thermodynamics of cooling system design, in order to be sure that sufficient cooling air reaches and moves by the part in question. You may be thinking, "But I don't have anything to do with the cooling of the parts." That may be true, but the impact of cooling air will directly affect your design—be it an ASIC (application-specific integrated circuit), a printed-circuit

---

2. This temperature gradient exists because of the thermal impedance of the material that forms the package that houses the chip inside the IC.

3. This subject is outside the scope of this book, but can be found in books on IC technology.

board, a power supply, a system component such as a disk drive, or a complete system. You need to be concerned, because the reliability of the design depends to a large degree on the availability of cooling air to maintain your product within its operating temperature range. And the reliability of your product hangs in the balance.

Also related to this temperature consideration is the subject of thermal stress, which you will find more thoroughly investigated in Chapter 9. Other environmental effects that your product will be subjected to would include humidity, vibration and shock, and altitude. Let's look at each of these, one at a time.

Effects of Humidity

Ever wonder why the military was so concerned about "conformal coating" on printed-circuit boards? Or why they are worried about the effects of humidity? After all, isn't the environment of a computer room humidity-controlled?

The problem, of course, is that *the product may not be used in a computer-room environment.* In the case of military products, it could happen to be a part of something that has to function in the steamy heat of a South American rainforest, or a jungle in Southeast Asia. There, high-humidity effects could very well mean the difference between the success or failure of your product. Or suppose your product is incorporated into a space-vehicle that forms an unmanned probe on the surface of Venus? (It can happen!)

Or on the other hand, suppose our product is to be used in the Sahara Desert. And to top it off, it is in the engine compartment of a vehicle being driven through that desert. Can you imagine the span of the humidity ranges it would have to operate in? Just suppose—if the engine were liquid cooled, and it overheated to the point of boiling over, even a wider range than you imagined could be imposed.

What we are advocating here is that you (as the design engineer) should be concerned about conditions that may be *beyond* the most horrible you could have possibly imagined. Do not assume for a single moment that your product will find itself in a protected, controlled environment. I can almost guarantee that it will not. So to prevent embarrassment and possible monetary loss due to oversight on your part, design to the worst conditions your product can possibly come up against. It really doesn't cost that much more, and it certainly will help you to sleep nights!

To help you to imagine some of these "worst" conditions, we will list a few here:

- Desert conditions, with low humidity ($\leq 10\%$) and high temperature ($\geq 100°F$)
- High altitude ($>10,000$ ft, with accompanying low ambient-air pressure)
- Cold ($\leq -40°F$), with both high and low humidity ($\leq 10\%$ to $\geq 90\%$)
- Space conditions (vacuum, temperatures as low as $\leq 100°F$, humidity as low as $\leq 10\%$) or the opposite, vacuum with high temperatures of $\geq 200°F$, and no humidity (humidity $\leq 10\%$).

If you were called on to design to space-conditions requirements, how would you guarantee that your product will operate in the above harsh condition specifications?

The only way is to design to those specifications and test, test, test—at those extremes until you have sufficient background data and proof that your product survives those conditions without a single hiccup for the required length of time. Don't make final decisions based on extrapolations, assumptions, hope, suppositions, hearsay, or whatever. Instead, you should be sure—investigate . . . run tests. Remember the references to Murphy's Law in previous chapters. Take the pessimistic viewpoint and design for the worst possibilities, rather than being an optimist (or worse yet, an "ostrich" and bury your head in the sand, so to speak).

For instance, let's say we are designing a piece of space equipment that embodies an embedded computer system. It is a piece of military equipment used for national defense. What are some of the parameters of the requirements or the environmental specifications we might be required to meet? We will begin with a hypothetical situation in order to answer this, and will carry this hypothetical case to its ultimate conclusion.

Hypothetical Requirements

The design specification reads:

The system shall be so designed as to operate within and have no susceptibility to the following environmental conditions:

1. Vibration of the following limits:
   a. Frequency span and amplitude: 1 cycle to 17,500 cycles per second at ± 1-in. excursions at 3 G
2. EMI of 0.001–1000 Gauss at 3 in., 30–1200 Hz
3. RFI of 100μV field strength at 1 m, of frequencies of 10 kHz–100 MHz
4. Temperature of −100 to +240°F, in combination with:
   a. Humidity of 10–95%
   b. Ambient air pressure of 7.0–35.0 in. of water
5. Operating input voltage range: 25.0–31.9 V DC with ±100-mV ripple
6. Maximum current drain and dissipation of 700 mA and 22.33 W dissipation

Pretty stiff requirements, eh? . . . Of course, this list of specifications is totally fictional. But, given these requirements and specifications, what techniques would the design engineer be required to use to meet them?

In order to get a better idea of the scope of this problem, let's take each requirement and analyze it, with an eye toward design specifications. First we would need to draw up a set of design specifications of our own that reflect the requirements as specified in the design requirements document. Our design specs would look something like this.

Design Specs

The system shall withstand the following with no failures, for the length of time specified:

- Vibration of the following limits:
  a. Frequency span and amplitude: 1-cycle to 17,500 cycles per second at ±1-in. excursions @ 3 G
  b. Length of time under vibration: 3 minutes

This requirement can be translated directly into mechanical antivibrational structural requirements specifications. This set of requirements then becomes an input to the mechanical layout and packaging engineer.

Many different techniques are applicable to this type of specification, such as "potting"[4] the circuit with a solid-walled enclosure with glass-bead insulation potted in place, or a plastic potting material-filled enclosure, or foam-filled milled-out box, to name a few. Other and better mechanical means of meeting this specification will probably occur to those who are experienced in this pursuit.

However, there are several considerations that will require input from you, as the design electronics engineer. Here, the mechanical or packaging engineers will need your inputs as to what detrimental effects the materials they might ordinarily select for the mechanical integrity of a unit might have on your design.

For instance, what are your heat-dissipation limits or requirements? (See sections on temperature and humidity/pressure requirements, which follow.) You should have direct input here, which will drive the material selection as much as the next question: What frequencies of signals are present in the device that might be attenuated by the wrong materials? You should consider the attenuation of signals by the materials selected anyway—not only for the reason stated above, but also because the amount of signal attenuation affects another important consideration.

EMI/RFI Requirements

- EMI 0.001–1000 Gauss @ 3 in., 30–1200 Hz
- RFI 100-μV field strength at 1 m, 10 kHz–100 MHz

4. Potting is the enclosure of a circuit with a material impervious to moisture and/or other environmental concerns

These two EMI/RFI requirements may be taken as a single specification, since they are both electromagnetic in nature. It becomes immediately obvious that the requirements call for some sort of shielding. But it may also mean that *two or more different types of shielding* may be required to meet the specifications, since it is true that varying *magnetic forces* (herein referred to as "EMI") exert a different effect upon electronics than does *emitted RFI*. However, the shielding for *both* might conceivably be combined into one material. Or it may mean that shielding can be incorporated as part of the packaging material spoken of earlier. If I were the design engineer in this case, I might start with a Mu-metal shield for magnetic-effects shielding, which, with proper grounding, could also be adapted to shield out RFI.

I suggest you get actual product information from the manufacturer—by simply sending for it. Or give them a call and explain your requirements . . . and let them give you a hint as to what part might best meet your needs. At any rate, equip yourself with the information you need to make an intelligent choice.

Temperature and Humidity/Pressure Requirements

- Temperature: −100 to +240°F, in combination with:
  a. Humidity: 10–95%
  b. Ambient pressure: 7.0–35.0 in. water

These requirements can all be considered together and handled as a single requirement, but must be considered when selecting components with which to construct the system. The temperature specifications of each individual component must be thoroughly investigated and considered.

The tendencies toward corona buildup must be a prime consideration, if voltages and frequencies are within the realm of corona-discharge phenomenon[5] at the ambient air pressures

---

5. *Corona discharge* is the tendency of a high AC voltage at fairly high frequencies to form a "corona" of highly ionized gas around the most pointed part of the circuit, (usually experienced at lowered ambient pressures).

(or rather, lack of pressure) expected in outer space. This phenomenon can cause a transmitter to fail completely in space. In fact, I have more than once been a member of a project team faced with this problem.

Voltage and Current Requirements

- Operating input voltage range: 25.0–31.9 V DC with ≤ ±100-mV ripple
- Maximum current drain of 700 mA and 22.33 W maximum dissipation (31.9 V × 0.7 A).

These requirements can be taken together, since they are akin in the respect that operating voltage and dissipation are related. They are a direct influence on the first and second requirements, for the reasons given there.

Again, the electronics design engineer must select the components with these requirements in mind. The entire design will hinge on the components selected to meet these specifications. Therefore, you should begin by ordering component catalogs from every manufacturer who makes parts that might even come close to meeting your requirements.

This helps by allowing you to rule out those components that do not meet the requirements, and even going so far as specifying that they *not* be used, if there is a possibility that later on the purchasing department may decide to try them because they are cheaper or are more readily available.

Remember Murphy's Laws

Nothing puts a design engineer at greater risk than to design with one manufacturer's parts that meet the requirements perfectly (and only those of that manufacturer). You may be doing this only to find out, further down the line, that the manufacturing department is putting another manufacturer's parts in the product because they are cheaper or more readily available, but the parts don't meet the requirements and are causing problems. Head off this type of problem in the beginning by specifying parts from those manufacturers only, and exact part numbers that you have tested and know will fill the bill. Further, in case the product is government-contract-related, be sure that the components are all on the "MIL-SPEC approved" list.

The foregoing examples were, of course, a fairly stiff set of specifications for a very extreme environment and for the most part will be beyond nearly all requirements you might run across . . . but don't bet on it. They are fairly close to those requirements imposed by spacecraft environments.

By those standards, all other environmental requirements here on earth may seem mundane. For some computer system environments, however, they may not be that far out. You may not find these particular requirements all together at one time, or even on one project. But design to meet them, and there will never be an environmental problem with your system. Of that you can be certain.

In these days of reduced-cost design requirements, you may be pressured into compromising a design to make it cheaper to produce. Don't do it. In some projects, you may be under pressure to order the cheaper or more expedient solution to a problem . . . Don't do it. If a certain level of environmental stress is expected, design to exceed that level by at least 10%. You are the design engineer—the product will be no better than your design, or the products you specify as components to be used in its manufacture. You will have to live with the final results. Don't compromise.

But then don't go overboard, either. Just because a component costs more than another doesn't guarantee that it is better. Do your job completely. Test *both* components at the worst conditions expected—then go 10% beyond the extremes of the range specified. You'll be glad you did. You will learn things you never expected.

On the other hand, if you are the customer and are buying components or parts from another company, don't accept a product until it has been tested by your own facilities and proved to pass the specifications claimed for it by the manufacturer. And by the way, don't test on a sample of one. Do your job thoroughly, and test at least 10 pieces of any product. Compile a set of data for those 10 and plot the results. Find the extremes of the batch. Then find the average or mean of the batch, and check it against random samples every so often. Design for that mean as the center of your tolerances, and specify the extremes as the acceptable limits for incoming parts. Continue to check your incoming parts against the acceptable tolerance limits you have set up.

By doing this, you have a credible center-point of the product you are testing, around which you can expect the rest to fall ($\pm$). By specifying this center-point or mean as the nominal and then setting your tolerances from that and designing for that nominal, you should never receive any surprises (as long as the manufacturer doesn't change anything—which you can bet will happen, at the most inopportune time).

Research and development (R & D) can be defined as a series of proofs that Murphy's Law does, indeed, exist (as we discussed previously). *Whatever can go wrong, will go wrong.* Given a chance, the worst will exceed any possible stretch of the imagination. Expect it, and there will be no surprises. If you get the idea here that I'm a pessimist, you're right. I am. I don't expect the best and then get stuck with reality. I expect the worst, and hope for the best. This way, I'm pleasantly surprised if the results are better than I had hoped, but I'm not totally devastated if they turn out to be worse.

And while we're on this subject (even though it may be slightly aside from the point of this chapter), I should point out that if our industry in this country is to succeed in these times of stiff competition, the object of everything we do should be reliability and quality. Now, this may sound like a cliché (and well it may be), but the one single attribute that offsets cost or nearly any other consideration of a product is *quality*. And here, "quality" could be defined as the triad—reliability, maintainability, and testability.

With that thought in mind, design for the worst, hope for the best, and expect to make a compromise. But make the trade-offs favor quality. Make the product you work on one you can be proud of. Sure, it may cost a tiny bit more during the Research and Development phase. But all things considered and all things said and done, it is the reliability, maintainability, and ease of maintenance that really makes one product better than another. And if your company's product is to succeed against the competition, it *has* to be better. *Design* it to be better. Use better parts. Make it more reliable than it *has* to be. Then, friend, I guarantee you, it will *sell* better!

Environmental effects may not be your worst enemy, but they are far and away ahead of whatever is in second place! ... And now that I have said my piece, we can get on with the rest of this chapter.

More on ESD

Right now, we will take another, more detailed look at our old friend—ESD. I am bringing it up again here to reenforce what we already covered and to add some additional information. For instance, we have not yet detailed all we can about testing for ESD and the results it can have on a system.

As we found in a previous chapter, not too long ago, ESD testing was left totally up to the engineer's personal resourcefulness. The engineer had to figure out what parameters needed to be tested, and how to accomplish the test itself. In those days, test setups had to be devised that were hopefully repeatable and would give solid, repeatable results. No rules existed that would accomplish this. It was every man for himself. Usually all you could do was test to ridiculous voltage levels (which in turn, brought with it extremely high current levels—way above that required for meaningful ESD tests). Then, if the circuit survived, it would withstand anything thrown at it.

But this required extreme preventive measures that were expensive and beyond actual requirements. Finally, the fact that some kind of standard was needed sunk its way into the consciousness of those who could do something about it. The American National Standards Institute (ANSI) led off the standards here in America, while the European Computer Manufacturers Association (ECMA) and the International Electrotechnical Commission (IEC) arrived at standards and guidelines for ESD testing.

The IEC 801-2 Draft 4 released in 1988 defined setups and methodology for ESD Testing. I *know* the methodology works, because by some strange coincidence, it is almost exactly the same approach I used in 1983 for ESD tests. And in every case, my tests were repeatable.

I knew I needed a good solid ground reference and some sort of screening to protect the rest of the building (which also housed the software development area and the manufacturing area). Without some precautions, this would have been very foolhardy, indeed. The first thing I did was lay a large (6 ft square) copper sheet directly on the concrete floor of the screened-off area I was to use for tests. This copper sheet was then connected by a short No. 00 braided wire to the facility ground. This became our ground reference.

We borrowed a test tool (called the "Zapper"™) from a supplier for a series of tests. The "Zapper" (as mentioned in Chapter 3) was used for RFI testing. The ESD tests were accomplished in a slightly different manner, as will be explained.

I was already very familiar with ESD, having spent a period early in my career "playing" with high voltage AC, in both the high and low frequency ranges. I once had a high-voltage lab where I built and tested such fun devices as the famous "Wilmhurst machine," Leyden jars, a surplus 27-MHz "RF Diathermy[6] machine from a hospital, and various sized Tesla coils—as well as hand-operated "telephone generators" from old rural telephones. For purposes of these tests, these and other more common equipment, such as neon transformers, were pressed into service.

The tests I did in those days were strictly from curiosity, and were not as scientific as they could have been. But it was all good experience that stood in good stead for the job at hand. I guess I could throw in a moral to this little story, and say that any opportunity to learn *anything* will eventually pay off in some way or another—but we don't do things like that in this book . . . (??)

At any rate, back to our story . . .

The device under test was plugged into an AC outlet that was connected through a line conditioner and filtered back to facility power. In this case, the line conditioner was to prevent

6. Diathermy is the technology of inducing artificial fever into portions of the body by RF induction.

feedback of spikes and surges out through the AC power (known as *conducted RFI emissions*) to the rest of the building, instead of the other way around. The device under test was placed on a table in the middle of the "screen room," which was 8 ft by 8 ft, with a 7-ft ceiling.

The source voltage was generated by means of a piece of commercial gear very similar to the flash circuit used on older cameras. The lead of the output could be brought within the required proximity of the device under test (DUT) for air-discharge tests in the vicinity of the portion to be tested. Thus, a high voltage source of DC was made available to be released to the unit under test, with a resulting spark. This single-discharge current spark brought forth fairly repeatable voltage waveforms that could be used for ESD testing. The output was a very high rise-time waveform, of many thousands volts.

In this way, induced voltage surges could be tested for. The output could be also fed to the device under test utilizing a direct connection through the finite impedance of an *RC* network. The air discharge appeared as a bluish-purple-colored arc, accompanied by a very sharp snap. Both methods yielded very good results in ESD testing, at a reasonable price. Another advantage of this setup was that an ESD discharge could be "generated" at will, with each discharge directed at a particular part of the device being tested.

A major disadvantage of this setup was that when utilizing this equipment, destructive testing was the only way to achieve useful results. But useful results were achieved—and the experience gained from these tests has been applied in many ways. It proved the worth of ESD-proofing the product, and therefore greatly improved the reliability of the equipment we were putting out in the field. In addition, along with other tests, it provided a basis for comparison of various peripheral products we considered adding to the system.

Failure diagnosis, in combination with the use of postmortem failure-mode investigation, was a very valuable source of information. It provided early insight into both the modes of failure and the exact mechanism that caused it.

Yes, this was a very primitive attempt at ESD testing, but it worked very well . . . Of course, I would be the first to admit it might not be acceptable these days. This primitive approach has long since been supplanted by very sophisticated test equipment (such as that built by and available from such sources as KeyTek). To get further information on this subject, consult a copy of one of the magazines that cover this subject such as *EE, Test & Measurement, World,* and others and fill out one of the "bingo cards" included therein. Or contact a company that specializes in this pursuit directly.

Do not underestimate the worth of ESD testing. To be certain that ESD won't be a problem, you really should run your own tests . . . even though the manufacturer may claim to have run these tests and provides data. The fact is, their data may be taken utilizing tests that are conducted under circumstances totally different from those you are most interested in. Or the data provided may be from a slightly different model, and is therefore of questionable value. Believe me, testing will prove to be worth its cost. Or on the other side of the coin, if tests are *not* run you may end up losing a lot more than the tests would have cost. (Besides, its a lot of fun!)

You will discover weaknesses you never imagined could be there. The testing you do now will prevent very embarrassing moments down the road. Besides, you will learn more than you could ever imagine. If you are anything like me, the investigative nature of this type of test is really a very enjoyable experience.

But on the other hand, don't do as the EMI test manager of one company I visited was doing . . . he had set up a test site on the second floor of the company's very nice building for EMI/RFI tests. The test site was complete with a turntable and antenna and was in close proximity to other offices, equipment, etc. I doubt seriously if they could get usable data from this test site—even good emission measurements. And I know for a fact they would have had real problems when trying to do any EMI or RFI susceptibility testing (if they even tested for EMI/RFI susceptibility at all). For if you aim a fairly high-power source of RFI at a test setup out in an open bay of a building with no way to contain it, every computer in the building will get the results!

When asked if they had considered using a screen room, he

said they had thought about it, but were advised that a screen room did not give repeatable results. But the point is, how could they get repeatability with the setup they had?

If your company does not wish to run these tests themselves, just take a unit off the line and run a demonstration for management. If this doesn't do the trick, there are small and large companies who provide a testing service at reasonable costs. These firms will test to their own specs, or will test to your particular specifications. Either way, the tests are not only *beneficial*, they are *indispensable*.

CHAPTER 9

Thermal Shock

In this chapter, I am going to discuss something that has been the subject of considerable debate among computer people for some time . . . and will probably be more so with the readers of this book. And that is the question, "Should computers ever be shut down and turned off unless absolutely necessary?" I am sure this will stir an opinion within you already, but wait . . .

Going a step further, I am going to talk about a slightly different, but related subject that there has also been a considerable amount of controversy about . . . the question of whether or not cooling air should continue to be supplied after the system power is removed.

The reason for all the controversy centers around the effect "accelerated cool-down or warm-up" has on components. The procedure of leaving cooling air on after power has been shut down has always been in question—such as in the typical case where the components have been running at normal operating temperatures with power on for an extended period of time and then are *suddenly* cooled down by a continuing cooling airstream. The first question therefore deals with whether that air stream should continue after the power has been removed. Also questionable is the exact opposite case—where parts that have been cooled for an extended period are suddenly powered up and quickly reach operating temperatures (unavoidable if the system power has been turned off).

The resulting effect of both these procedures is known as "thermal shock"[1] and is responsible for many failures in the field.

Look at it this way: on initial power-up, a computer's components are brought up in temperature from "dead cold" to the final operating temperature they will attain during normal operation. The upward trend of temperature change here tends to be fairly slow rather than instantaneous, due to the components being cooled by airflow or air-conditioning during the warm-up period. The reverse is actually a lot worse, where a component has been operating for an extended period at a certain temperature, and is suddenly powered off and cooled rapidly by a stream of cooling air that continues after the power has been removed. Here, the *rate* of change in temperature is much more severe. You may argue . . . "But the part is not under power in this situation." But in either of these cases, the part is subjected to several different kinds of stress—both mechanical and electrical in nature—all at the same time even though no power is applied at that particular moment.

The term "thermal stress" points out that ICs and other components experience physical stress as a result of sudden changes in temperature. Actually, the stresses imposed in the above situations include several different kinds:

- *Mechanical stress* on internal structure of the device itself (caused by unequal expansion and/or contraction of the various materials used in the device)
- *Electrical stress* caused by sudden removal or application of voltages to a device
- *Physical stress* on an atomic level due to sudden changes in temperature, which affect the actual composition or charge distribution of the materials incorporated in the structure of the device itself

Thermal shock is present no matter what length of time is involved in the change of temperature (i.e., rate of temperature

---

1. Thermal shock is a very big factor in the failure of computer systems (see Glossary).

change). But it becomes worse as the time span becomes shorter (or a higher rate of change), and/or as the *temperature span* involved becomes wider (because this also increases the rate of change between extremes). As we shall see (and you probably already know), these two factors combine to wreak havoc in a number of ways.

Effects of Thermal Stress

Some of the purely mechanical effects seen from thermal stress include

- ICs loosening and "backing out" of the socket they occupy in a printed-circuit board
- The breaking of pins or solder joints
- Drying out of certain components such as paper and electrolytic capacitors[2]
- The lengthening and/or shortening of leads due to temperature changes with "metal-fatigue" breakage as a result
- Broken bonding wires internal to the IC

. . . and so on.

Intermittents Caused by Thermal Stress

The "backing out" of ICs from sockets seems a strange phenomenon, until you understand it. In case you don't, the mechanism behind what happens is this: the part is originally installed snug and tight in a socket while the board is dead cold. When power is applied, the board begins to heat up to operating temperature. As this happens, the pins of ICs in sockets serve as the thermal path for internal heat to reach the cooler sockets and solder joints. As the IC heats up, the pins themselves expand lengthwise slightly because of thermal expansion from the increase in their temperature. This acts to push the IC away from the socket ever so slightly.

---

2. This may become much less of a problem in surface-mount technology (SMT), which utilizes more "chip"-type components, with shorter and stouter leads.

Then eventually, the power is removed and the board begins to cool . . . either by itself with convection cooling or by the forced cooling air remaining on after power is removed (which causes much more rapid temperature variation inside the part). As the part cools, it begins to contract slightly, which tends to pull the leads or pins out of the mating socket ever so slightly. This process continues with each heating up and cooling down, causing the pins to withdraw a little more each time until eventually a pin on a part somewhere on the board no longer makes good contact. At this time, a "thermal-intermittent" rears its ugly head. If it is not recognized, found, and fixed, it soon becomes a "solid failure." The biggest problem, however, is that as the temperature changes, this bad connection is intermittent with temperature—making it very hard to find.

This is the mechanism behind this particular kind of mechanical failure of a circuit from thermal shock. And believe me, this type of failure is well known by the CEs (customer engineers) who service customers' computer systems regularly. The CEs are the poor folks who have to diagnose this kind of problem and then find it and *fix* it under the worst kind of conditions. If you doubt the value of nipping this failure in the bud, just ask one of the *CEs* if *they* would like the design engineer to design this kind of problem out of the product!

Start-up Failures Due to Thermal Stress

Another failure mode that shows up suddenly after a system has been running continuously for a long period of time is *power-up failure*. This mode is actually a latent problem that simply shows up (most often) after the system has cooled down and is then suddenly brought back up under power. The typical scenario is usually as follows: A system has operated continuously for a long period of time without failure, and is then shut down for some reason or other—long enough that it is stone cold—then it is powered back up.

At some point during the power-up cycle, it suddenly becomes painfully obvious to all concerned that this computer is not going to "come back up." This is made clear to the operator in many different ways, but they all boil down to the same thing—the system is "down for the count." The system has exhibited a hardware failure between the time it was powered down and when it was powered back up. Another start-up failure.

The failure mechanism is usually pretty much the same—and that's true even though the type of system *(and sometimes the failure's symptoms)* may be totally different.

The Start-up Failure Mechanism

This failure mechanism is fairly simple, and works as follows. As long as power remains on in a system, the parts remain at very nearly the same temperature. If its internally generated heat is removed at a rate equal to that at which it is generated, it reaches an eventual temperature equilibrium—which is hopefully within its operating range, for longevity. It may continue to operate without incident for months—even years. But when that part is allowed to cool by powering it down, it soon becomes dead cold because no heat is being generated to offset that being removed.

If you apply power at this point, resultant stresses will be generated within the innermost structure of the part itself—which are destructive in nature. These may be due to a number of things, among which are these possibilities:

1. Instantaneous voltage changes at "turn-on" (such as voltage differentials and transients) that bias the diodes, transistors, etc., inside an IC very slightly the wrong way or go beyond its breakdown voltage[3] (a form of electrical stress)
2. High instantaneous *currents*, which are beyond the part's current capacity, occur and result in "blowing it out"—caused by improper bias voltages during start-up (this failure is also from electrical stress)
3. "Punch-through" occurs as a result of a voltage spike during start-up whose voltage is higher than the breakdown-voltage rating of the part in question (again, electrical stress)

---

3. Sometimes known as "reverse breakdown" or "punch-through" (see # 3 and 4).

4. "Reverse breakdown" of a diode or transistor occurs during start-up, due to a reverse-voltage spike (more electrical stress)
5. Mechanical contraction from rapid cooling after power shutdown and expansion during power-up from heating up (mechanical stress)
6. The repetitive expansion and contraction causes a part to "back out" of its socket (mechanical stress)
7. A reduction in internal resistance between two portions of an ICs circuit, causing leakage or breakdown (physical stress)

Diagnosing Failures

Failure diagnosis is an ongoing effort at all the better component manufacturers and results in most failure modes being pinpointed and designed out by the manufacturer before they cause widespread failures. But sometimes, a cold start-up after cooling down between long periods of running at normal operating temperature generates stresses not duplicated in any other way.

"Accelerated cool-down" (from the air-conditioning airflow continuing after the input power is turned off) can be destructive, if not planned for in the design of the system from the start. Some system manufacturers have seemingly managed to design the airflow such that continued cooling air supply does not cause destructive thermal shock. But for the most part, I think you'll find that it is considered detrimental to the hardware. In fact, lots of IC manufacturers will specify a cool-down rate not to be exceeded for their products.

If you doubt or are unsure of the effects of thermal shock on a circuit's components, you have only to put an operating circuit into an environmental chamber and run these tests.

1. After putting the circuit into an environmental chamber *without* the board's power being applied, take the chamber down to −40°F. After the part has "soaked" at this temperature for some 2 hours or so, suddenly apply the power. Run comprehensive tests on the part while in this

condition, and watch results during the ensuing "warm-up period."
2. Then the reverse—after putting the circuit into the chamber, take the temperature up to $-140°F$ *with board power applied.* Allow the circuit to "bake" for an hour or two, then turn off the power and take the environmental chamber down in temperature to $-40°F$. Then let it soak there for about 2 hours—and then suddenly apply power. Again, run tests during the warm-up and carefully note the results.

If the circuit "comes up" without a failure, you have a very hardy circuit design, indeed.

What I am saying here is not that we want to destroy the circuit by unrealistic thermal stress—for the conditions described are not unrealistic. On the contrary, we are simply duplicating conditions that the circuit could very easily encounter in actual usage . . . conditions that it should withstand without failure. In fact, I would not release a design that could not survive this type of treatment, at a minimum, . . . repeatedly.

As was pointed out in the chapter on burn-in . . . when a part has been thermally cycled several times during the burn-in period and continues to operate, chances are it will continue to operate for an extended period of time in actual use. However, there eventually comes a time in the part's life span when thermal stresses will take their toll. Thermal shock accelerates the normal aging process, and will shorten the life span of any part by a very significant amount.

What to Do?

So, what is the bottom line? What can we do to overcome the possibility that thermal shock will lead to the early failure of our product?

Again, the first line of defense is to design with these failure modes in mind—in the following ways:

1. Use only sockets that are proven to prevent parts from backing out and/or losing contact under thermal changes, or else (better yet), do not use sockets at all. Instead, wherever possible, solder the parts in directly. (You'll save the cost of sockets, anyway.)
2. Buy or specify only the most reliable parts, and be sure they get tested thoroughly, while at the same time being thermally cycled up and down in temperature during burn-in.
3. Look for possibilities of failure under thermal stress everywhere in your product.
4. Do thorough environmental-chamber testing of your product before you bless the design and pass it on to manufacturing.
5. Get together with the manufacturing engineer(s) and have them pass on the design. (A design review is always a good idea, anyway, because there may be something very devastating in its consequences that you have entirely overlooked.) In fact, there is a trend toward getting manufacturing and sustaining engineering involved early in the design cycle . . . which I think is a very good idea—their inputs may save you lots of embarrassment later.

In addition, do as I suggested earlier—take the circuit up in temperature to +140°F with power on and soak it for 2 hours. Then turn the power off, and take the circuit under test down in temperature to −40°F—soak it for 2 more hours, and then turn the power back on. Repeat this procedure for more than one or two cycles. If your product survives, it will probably survive out in the field. If not, find the failure mode and fix it.

The next thing to do, of course (and this is actually a very good idea, anyway), is to write a set of specifications that will *prevent or preclude* the conditions that you know your product cannot possibly withstand. As a matter of fact, this should be done even if your product survives all the tests. (More information on this subject in the next chapter.)

The best of all worlds (as mentioned before), is to work with the mechanical, packaging, and manufacturing engineers during the design of the product to make sure that the design will be on target in all these areas. The early design stage is the time to do this—it will be to no one's advantage to find out the design cannot be manufactured to meet specifications at a later date, after the design is complete.

Besides simply making friends with manufacturing, this technique will help to guarantee a truly reliable (and manufacturable) design—with very little chance there will be inherent, unforeseen, designed-in problems. This may not make you the most popular engineer, but it certainly won't hurt your reputation for reliability.

I think I should warn you however, that the company's main interest may not necessarily be in the product's reliability, these days (but it *should* be)! But it's actually a part of your job to convince them that product reliability (or the lack of it) will make the difference in the useful life of that product.

Besides, who wants to be plagued with a design that is eternally popping up with problems . . . requiring you to constantly "put out fires" instead of doing productive, enjoyable engineering? I know from experience that it is a much more satisfying feeling, knowing your product is quality engineered for reliability. And contrary to the present belief that cheaper is better, the word gets around. Those who buy an inferior product (and get "stung") talk about it. A cheap solution to thermal shock may be very expensive in the long run.

Actually, thermal shock is (in a way) a reliability factor beyond the manufacturer's control. But if it causes failure *within the warranty period*, thermal shock can be a very expensive destructive force, indeed. But that does not mean there is nothing we, (as the designer) can do about it.

By specifying operating conditions that will result in operation within safe parameters—and assuring (through training and user's manuals) that parts are not being inadvertently overstressed during routine maintenance and operation—you have greatly reduced the risk. Through working with the mechanical engineer and manufacturing to achieve the required packaging and thermal design, you have reduced the risks even further.

Thermal Dissipation Problems

Since thermal dissipation is applicable to the subject of this chapter, it will be discussed here. Because it is within a design engineer's control —his or her inputs will help determine the packaging requirements—thermal dissipation tests should be run on the final design. And they should be conducted with as near the final packaging as possible. Look for hot spots caused by cooling-air "stagnation." Make sure that all parts of the circuit are getting sufficient cooling air. For in actual practice, cooling air is the "life blood" of electronic circuits. High thermal operating temperatures mean shorter operating life.

Don't rely on mere calculations or the manufacturer's specifications to arrive at thermal dissipation figures. Do an actual current-drain test of the entire assembly. Then use the resulting power-consumption figures to arrive at thermal dissipation for your product (assuming it is a board-level or higher product).

The next step is to transfer this information to those who really need it to make an intelligent decision about packaging. And give them "worst-case" figures . . . it may be higher than they want to hear, but don't let that make you lower your figures just to please them. You have to remember that reliability hangs in the balance, and lack of reliability never sold anything.

This is extremely important, in the case of an entire system, because the mechanical design of a card cage inside an enclosure can make all the difference in the world . . . because it may be related to eventual operating temperatures. Because, even if your design is absolutely perfect in every way except for the packaging, it can be a total failure as a final product. The *"thermal packaging"* determines how the heat will be dissipated. If the circuits are too confined to allow proper air circulation, the result will be overheating—with obvious consequences.

By "doing your homework" and running current-drain tests, inrush current tests, thermal dissipation tests, and the previously detailed "temperature-extreme" environmental tests, you have all but guaranteed success of your design in the field.

By the same token, if all the facets of the final functional requirements of the system are not examined early in the design, Murphy's Law has been given a free ticket, and will "do you in."

Other Effects of Thermal Shock

One of the most aggravating results from thermal shock is the thermal intermittent problem we mentioned earlier. It is aggravating because, being intermittent, it is extremely unpredictable and hard to find. I've seen technicians spend literally hours chasing down a thermal-intermittent on a large complicated PC board that is a part of a large system.

Most technicians can (once the problem has been isolated to a particular board), rapidly find the thermal-intermittent *component* on that board by using a cooling spray or heat gun. But the act of isolating the problem to a particular board in a large system can take hours.

The biggest problem with thermal intermittents is the fact that they can escape detection—or (worse yet) they may not become a problem until the product is placed in service. But by that time, they can then be very expensive . . . in terms of both money and the black-eye they give the company's product reliability as a result. And customers seem to take a dim view of a customer engineer having to spend hours troubleshooting a thermal-intermittent.

On the other hand, most thermal-shock victims start out as an intermittent problem—especially those that result from the mechanical aspects of thermal shock, as explained earlier. The results are usually perplexing, because they don't follow a predictable and repeatable pattern of behavior. They may act totally differently, two different times in a row. And because the point at which they become intermittent is related to temperature—instead of a particular routine or action on the part of a program—they do not display a logical pattern.

The moral to this story is that if you don't want to become familiar with the failure modes first-hand, *in the field,* then become familiar with them in the lab, by running tests and observing the results.

CHAPTER
10

Preventive Design Techniques

For the remainder of this book, we will for the most part discuss the "total solution" to the ESD/RFI/EMI problem.

To do this, it is necessary that we cover the individual board level and work outward. However, in order to cover all of the board-level considerations—such as printed-circuit-board layout, bypassing and filtering, ground-plane design, etc., we would need an entire chapter for just the board. So we have done just that—we've given that subject a chapter of its own and you will find PC-board considerations (highly detailed) in Chapter 11.

Therefore, in this chapter we will concern ourselves with a closer look at other more common aspects, such as breadboarding, prototyping (where used), and our proven zzaap-proofing techniques.

Protyping and Breadboarding

To coin a phrase, "a design is never any better than its prototype." And believe me, I'm willing to go on record here as saying that I definitely advocate prototyping a design—because I am certain that (contrary to what some articles and ads I have read of late would have you believe)—even though a design is done using CAD/CAM (computer-assisted design/manufacturing) techniques, *it still should be prototyped!*

This is especially true if your design includes any analog circuits. I say this because I do not believe personally that there is any way an analog circuit can be modeled and simulated to the degree necessary to provide a high enough confidence level to start production without a prototype. (This includes even those fortunate enough to be able to afford the most sophisticated and expensive CAD/CAM systems we have today.) In my opinion, to produce an analog design without prototyping is nothing short of foolhardy. I am not even sure that a digital design could be done this way.

"But . . .," you say, "the CAD/CAM systems available today do away with the prototyping requirement." I'll go on record here as saying I would have to see some of the circuits I have worked with in the past designed using CAD/CAM with no prototyping—*and then work the first time*, before I would be convinced this will work every time. If a circuit contains *only digital circuitry*, there is a remote chance that you might be lucky enough to produce a working production model with no prototyping. But there is no way I would even *consider* trying it with any analog circuits. And this is even more true if the design includes analog circuitry such as A/D or D/A devices, audio, radio-frequency or modulator circuits. Why is that? Because analog circuitry is highly dependent upon layout, due to interaction of components, dependency upon good grounding techniques, and other considerations. Digital circuitry is much more forgiving.

However, on the other side of that coin, I must admit that some people I have known could take a perfectly good digital circuit design that works in prototype, and make it unmanufacturable due to the prototyping techniques they used. Hard to believe? I've seen it in action!

Case in Point

Years ago, when most circuits were analog (and those that *were* digital were built using discrete devices), I once worked with an engineer whose favorite approach was to build a breadboard of a digital circuit on one side of a sheet of copper-clad printed-circuit board. Now I realize this form of prototyping has long since gone by the wayside—but for purposes of this example, please bear with me. For it is more his lack of forethought and unique technique that I wish to exemplify here, than the state of materials or technology at that point.

Being a very thorough engineer (even if not a very accomplished technician) he would do all the necessary calculations, truth-tables, etc. that were required during that era to design his circuit. He would do this only after he had drawn up a rough schematic, and checked it over with a fine-tooth comb. He would then proceed to assemble the breadboard as follows.

This Is a Prototype?

Using the normal breadboarding technique in vogue for high-frequency circuits at the time, he would put each active component [transistor, op-amp (operational amplifier), FET (field-effect transistor), or whatever] upside-down on the board as he came to them—using standoffs where necessary or attaching them by the leads that were destined to be a ground to the copper-clad ground plane. (This breadboarding technique is sometimes called "dead-bugging" ICs on the board, due to the resemblance between an IC or transistor, with its legs up, to a dead bug). The advantage here was supposed to be that all parts were as close to the ground plane as it was possible to get them—giving them protection against airborne noise (RFI). The active parts were then joined by the resistors and capacitors, stretched between the active parts, and soldered.

He would not always shorten the leads of resistors and capacitors before attaching them, and would support them by connecting the leads in midair to other components. He formed a grid of components as he went along, tediously soldering each lead of a component to an IC pin or to the leads of other components, until his circuit resembled a large unwieldy spiderweb.

When this process was completed, and he had assembled the entire circuit, he began his testing. If he found he needed to change the value of a component such as a resistor or capacitor, he would not desolder and remove the part. Instead, he would either parallel it or connect another in series to reach the desired value that he thought the circuit required.

By the time he arrived at a half-way functioning circuit, he had a ball of components resembling a bird's nest (or a spider's web), with components and leads intertwined and connecting in every direction. If he finally got the circuit operating to his satisfaction, he would then jot the final values (obtained by calculation of parallel and series values) onto his schematic. He would then turn his design over to drafting to be laid out, drawn up into official drawings, and distributed to manufacturing.

He was very meticulous about making sure the schematic matched his circuit—except for one very important detail . . . *layout!* Needless to say, it was impossible to duplicate his design and get it to work. Because of all the stray and distributed capacitance of individual components to each other and ground—plus the added inductance of the long leads, his circuit was a nightmare to phase into production. Built on a printed-circuit board, with single components of the values he had indicated, the circuits simply would not work. And each time, either he or a sustaining engineer had to make revisions to the circuit in order to finally arrive at a workable circuit.

Needless to say, this approach would not be well received these days. (They weren't then, either.)

Even now, with modern circuit design and prototyping techniques being what they are, breadboards are still being built . . . some with methods similar to those used for years, although they are now much smaller and less time-consuming than even those built using the ubiquitous wire-wrap technique. But they, too, have a similar problem—that of the huge difference in how a breadboarded circuit behaves versus how it will behave in the eventual printed circuit [albeit surface-mount technology (SMT)] form.

Don't take this approach and risk having this happen to you. If you do breadboard, use a more or less final layout that can be easily duplicated, with as short leads as possible. Be sure that ground leads are truly in the places they will appear in the printed version of the board, and are not daisy-chained or randomly placed. Be certain that power coming onto the board does so through accepted ferrite or other RFI-suppressing devices. Use only the accepted and established techniques that have been proved to work. This is most important, because a mistake in the early design stages during breadboarding can cause many hours of redesign and/or gnashing of teeth and wringing of hands, later.

In places where the circuit is actually breadboarded, be sure to follow established, proven grounding and noise-suppression rules—and don't cut corners. There are enough things that can go wrong without giving Murphy's Law a free hand and contributing to them yourself. Design and build the way the final product will be, as nearly as possible. You will be glad you did.

Even if the circuit is a digital design utilizing all the latest technology and it is designed by means of the best of the CAD/CAM world, you should still subject the circuit to very close scrutiny, with RFI and EMI in mind because even the most sophisticated CAD/CAM cannot yet predict EMI/RFI patterns. And do this early . . . in the engineering-model or prototype stage if possible. Look at the frequencies (caused by extremely fast rise-times) being generated or conducted by the circuit versus lead length. See if you can discover a point at which the lead length equals the wavelength of one of the circuit's frequencies or its harmonics (both upper and lower).

As frequencies go up, these wavelengths will obviously become shorter and shorter, so that size of the required real-estate shrinks—but by the same token, shorter and shorter leads means they tend to resonate at higher and higher frequencies. In fact, it is highly probable that engineers involved in the design of high clock-speed reduced-instruction set computers (RISC) processor systems, memory systems, and other circuits including those utilizing many of the new application-specific integrated circuit (ASIC) devices, have learned they have to deal with this consideration on an everyday basis. They probably accept this as a given, and think nothing of it.

However, it is also within the realm of possibility that there are some who have been bitten by lead-length-related resonance and do not realize it—yet. For unless it makes itself known in a repeatable, predictable fashion, lead-length can be overlooked until the circuit is further along in its development cycle.

As a matter of fact the track's lead length and width, along with the thickness of the epoxy–glass board, enter into a very predictable relationship. These parameters (which contribute to a predictable behavior) can be plugged into a reactance formula and the result will give the particular "characteristic impedance" for that particular track's dimensions. Welcome to the world of *strip-line*. For that's really what we have here.

"Strip-line" is a special type of transmission line—one where the thickness and width of the copper path versus the thickness of the epoxy–glass board (or Teflon–glass board, ceramic substrate, or whatever) have the relationship shown in Figure 10.1. This relationship makes the assumption that the board is copper-clad on both sides, with one of those sides forming the paths or tracks and the other a ground plane.

The parameters shown in Figure 10.1 enter into a calculation that yields the characteristic impedance for those particular parameters. The formula is

$$Z_c = \frac{WT_t}{T_b} \int \frac{di}{dt}$$

where W = width of copper track
T_t = thickness of copper track
T_b = thickness of board
Z_c = characteristic impedance

The characteristic impedance of a strip of metal along a printed circuit on a substrate is important even to a digital designer, because it may result in an impedance mismatch between two parts of a circuit that is serious enough to cause

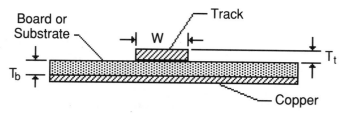

Figure 10.1 The "strip-line" parameters.

that track to become a transmitting "antenna," and radiate energy at that point in the form of RFI.

"Embedded" Processor Design

The design of embedded-processor systems is a special case that requires special care to prevent problems in the final manufactured product. This is true for several reasons. One of the biggest reasons is because most embedded-processor systems are "bare-bones" designs—having very little leeway in cost and size. This means that a design must meet requirements with as low a cost (in both parts and manufacturing costs) as possible. Another good reason is that embedded processors are now being used in environments that were heretofore impossible for a computer.

For example, a processor embedded in a microwave oven would have a tough time, indeed, without special thought in layout, filtering, and shielding. What about under the hood of a vehicle? Talk about hostile environments!

But any manufacturer here in the United States must be very cost-conscious in producing a marketable product. And an embedded processor is one way to greatly reduce production costs.

Back in the days when nearly all TV sets were still being manufactured here in the United States, a technique I saw used to reduce final product costs was: when a TV design was complete, cost analysts would remove selected components, one at a time, *until the TV no longer functioned.* At that point, those parts *without which the TV would still work,* were removed from the design.

Possibly a cost-saving technique, but no doubt a very poor reliability technique, I'm afraid. But be that as it may, this technique was used to keep the price of the TV sets down. But even this did not prevent the eventual take-over of the TV market by the Japanese.

In these days of SMT, CAD/CAM automated design and manufacturing, VLSI very-large-scale integration of circuitry, and ASICs, circuits are very easily designed and relegated to custom integrated circuits. If done properly, these miniaturization techniques help tremendously in improving printed-circuit-board real-estate utilization. Thus, they contribute to making everything smaller, all other considerations being equal.

Wherever possible, circuits should be designed to be fail-safe. At the very least, they should be capable of surviving under less-than-ideal conditions; i.e., they should be "hardened" against the more common environmentally induced failures. To do this requires that the design engineer be ever mindful of failure mechanisms that are linked to RFI, EMI, or ESD.

When the products resulting are as good as they can be, and you want to preclude failures induced by situations beyond your control, write specifications for your design that spell out environmental requirements. Make these requirements specific enough that they preclude the product being subjected to conditions *that you know they cannot survive!* In this way, your product will be assured an environment in which they will be happy . . . or else they will be excluded from warranty coverage.

The specifications for a product are best written by the engineer who designed it. This is because that engineer (assumably) has tested the product at all parameter limits, and knows what extremes the product will be able to survive and those that it will not.

However, spec limits should not be set from a sample of one. A reasonable number of production units must be tested to determine what the limits of survivability really are. These limits then must be expanded or reduced to cover all units tested that passed the tests. The final figure, minus a small "fudge-factor" to provide the "warm, fuzzy" will be the published limit.

There will be those who will totally and vocally disagree with this philosophy, of course. And that is their prerogative. Difference of opinion is one of the things that made this country great. And I would have it no other way. But at least, consider the things I have said here. Try them once. Then if you are not satisfied with the results, examine your implementation of the principles I have expounded on. Decide for yourself whether they have been given a fair trial. Leave nothing to chance. Keep good records . . . which, by the way, brings up a very important detail that many—way too many . . engineers neglect. That detail is documentation of everything they do. This includes original thoughts during the design and development process that will prove invaluable later, and the changes to the original design during its development—plus the "before-and-after" behavior that precipitated the changes. These notes should all be kept in a good engineering notebook. Some engineering managers I know judge the people working for them to some extent by the notebooks they keep.

Be that as it may, good engineers are known by the product they design. If they can continually design products that are economical to produce, and are reliable under adverse conditions, they will become known for them and soon can command the salary they deserve. (At that point, however, the "Peter Principle" usually takes over, and the engineer is promoted to a level of incompetence.)

Just kidding . . . quality at a reasonable price will always get the kudos. Put these into your product and you'll never be sorry.

Other Preventive Techniques

Grounding is another very important consideration, at both the PC board level and the complete assembly level. Grounding techniques can make or break a system design. They can also guarantee success for a design, by preventing the kind of problems associated with poor grounding . . . radiated EMI, RFI susceptibility, ESD damage, and so on.

Shielding sometimes is the only way to guarantee the protection of radio and television receivers against interference from a portion of a computer that is running at extremely-high-frequency clock signals, with extremely short rise-times. This is true because even with all the precautions talked about in this book, when the frequencies of processors get above a few megahertz, they will radiate energy regardless of precautions. These radiations can be contained within a shielded enclosure, and they will not escape—as long as they are not allowed to propagate by means of wires or printed traces on a board (conducted RFI). To completely isolate the circuit, then, it must not only be shielded, but all conductors into and out of the enclosure must be filtered, as well.

The best means of accomplishing this is probably to use ferrite beads. But if the conductor is printed on the board, an inductive and capacitive filter may be necessary. But all this is explained much more fully in Chapter 11.

For now, suffice to say that all precautions you can take toward proving your design by prototyping, testing, testing, and testing will pay off. Get familiar with all the anti-EMI/RFI techniques and devices. Learn to be able to predict when EMI/RFI will be a concern—*before* the design goes to manufacturing.

All effort in this direction cannot help but pay off.

And now, onward to board design . . .

CHAPTER
11

Printed-Circuit-Board Design and Layout

With all the discussion in foregoing chapters of things *not to do*, we really should sum up by getting into the *right things to do*. In taking a good look at the proper design and layout of a typical printed-circuit board, we hope to cover most of the mistakes that will cause grief in the production printed-circuit-board.

On the PC-board level, there are a multitude of things the engineer can do to help guarantee that their product will be totally reliable and will not be an offender in the radiated or conducted RFI/EMI department. This is true whether the board is the through-hole variety or a board utilizing surface-mount technology (SMT).

Let's assume for purposes of this discussion that we are going to design and lay out a printed-circuit board that contains both digital and—even worse—analog circuitry, and includes an audio transmitter and receiver on it. This audio transmitter sends audio to the receiver by means of modulation of an infrared beam.

Design Techniques

The first thought that comes to mind might be that analog and digital circuitry should not be mixed on the same board. And with good cause. But let's go even further and make a hypothetical supposition that our application requires a very small package, to be totally contained on one board. Unwise? . . . Possibly. Impossible? . . . Not really.

DC Power Supply to the Board

The first place we shall choose to begin with is the DC power supply entry point on the board itself. Whether this power connection is at an edge connector or is through a push-on cable connector, the technique is the same. Directly at the point where the connector pins are connected to the power distribution tracks on the board, there should be some kind of an arrangement that will allow the installation of a ferrite bead on each lead. The purpose of this, of course, is to prevent or stop conducted RFI/EMI (both onto and off of the board). The usual way this is done is to have through-holes at both the edge connector contact and the beginning of the track, with a single piece of uninsulated bus wire serving as a bridge. A ferrite bead is slipped over the piece of uninsulated bus wire, which is then soldered into the holes.

Following this, there should be a large-capacity-value capacitor (≥ 50 μF) connected between the power track and ground, serving as a bypass to ground for low-frequency disturbances. This capacitor should be as close as possible to the ferrite bead, with very short leads. Connected at almost the same point should be a smaller ceramic capacitor of about 0.1 or 0.01 μF to ground, to bypass the upper–middle frequencies to ground. This will guarantee clean DC onto the board.

All this sounds very complicated, but really isn't. A look at Figure 11.1 should help to clear it up.

Ground Plane and Ground Tracks

The ground should be distributed from the edge connector point to each of the other ground points on the board by as many individual *dedicated* ground traces as possible. Do not use "daisy-chained" ground traces. Also, every place on the board that is not actually occupied by other traces should be covered

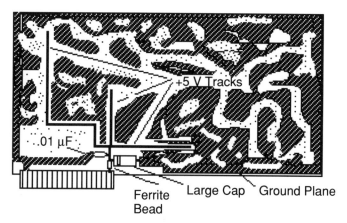

Figure 11.1 Power and ground distribution.

by the ground plane (except for required clearances between the ground plane and other tracks, of course). A ground plane should exist on both sides of the board.

Bypassing

The next thing to be assured of is that each IC or other active component on the board has its own private bypass capacitor connected between the component's power pin and ground. The bypass capacitor should be about a 0.01-μF ceramic, soldered in with as short leads as possible.

Segregation of Circuits

The next consideration is separation of *classes* of circuitry. By classes I mean, for example, analog versus digital. It is very important that any analog portions of the board be well shielded from digital circuits.

Within the analog class, there are several subclasses that should be separated from other circuits and/or shielded as well. They may also be worthy of some other peculiar considerations. For instance, let's assume that on a particular board, we have a transmitter circuit, a receiver circuit, and an oscillator, as well as some digital circuitry. A board with this many different signal types should be laid out very carefully, and is best approached in the manner described in the following section.

Layout Techniques

First of all, recent technological advances have given us components that make the above not only feasible, but practical as well. But in order to be assured that we are doing it properly, we must be extremely careful. For instance, let's start with power—getting it safely onto the board without carrying unwanted interference onto or off the board. This is easily done, but requires some insight or know-how (which you hopefully have, or will have gathered from this book).

Let's assume further that our power is +5 VDC, and we have calculated that approximately 350 mA will be required for the entire board. The first requirement we should bear in mind is a ground plane that encompasses any portion of the board not otherwise occupied by components or actual circuit traces (as pointed out earlier). This ground plane would hopefully be tied to a single-point ground. The positive side of the +5-V power should be distributed about the board by the use of parallel supply lines as follows.

1. The +5-V supply should enter the board through a ferrite bead feedthrough as stated earlier and shown in Figure 11.1. The ferrite bead should be chosen for a low-end cut-off frequency far below the lowest frequency encountered on the board.
2. A large-capacity value electrolytic capacitor should be provided for bypassing, immediately following the ferrite bead.
3. A 0.01-μF capacitor should be connected at the same point as the other bypass capacitor (as near as possible), with as short leads as possible to the nearest ground.
4. The 5 V should then be split and distributed to the various points on the board where it is needed.

These points can be seen more clearly in foregoing Figure 11.1.

Segregation of Circuit Types

Next, we must decide where each of the sections of the circuit are going to reside. This is a very important and critical part of the layout process, and should be done by a circuit design engineer who is experienced in both analog and digital—*not simply by a designer or draftsman*. This is because the experienced engineer will (hopefully) already be aware of the pitfalls of mixing receiver and transmitter circuits (not to mention analog and digital circuits) on the same board—where a layout draftsman might not.

As a rule of thumb, receiver circuitry should be the closest circuit to the power input. The transmitter circuitry should not draw current from the same point as the receiver, but should instead have its own feeder with separate bypass capacitors. The digital circuitry should be isolated to one portion of the board, bordered on *no more than two sides* by analog circuits. (This is a very important point.) And there should be a Faraday shield provided that is *taller than any other component* on the board, and which zig-zags as necessary around the borders between the three kinds of circuitry. This shield should be grounded at intervals of no more than every half inch. A nice choice for this job is the laminated strips available from several manufacturers. The strips provide several layers of metal insulated from each other, but rich in capacitance between the metal layers. If ground is connected to every other layer, this makes not only a very effective Faraday shield but also a filter for power. Just be careful not to connect two signal circuits to adjoining layers. (Cross-talk, remember?)

Any power going to one of the three circuit types should not traverse territory belonging to a different category of circuit, i.e., *power going to the receiver should not get there by way of crossing the transmitter area*. To do so would be begging for cross-talk or feedback that would be certain death to the receiver signal. Figure 11.2 shows the board with the three sections of our hypothetical circuit laid out following our rules of thumb.

For those who will be using the relatively new SMT, the PC board can be laid out so that a huge ground plane exists on one

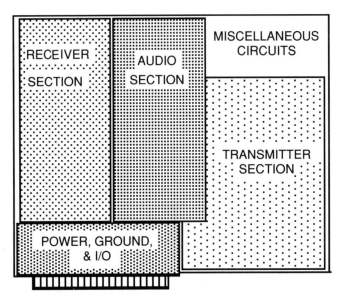

Figure 11.2 Layout of the three sections.

side of the board, with the components and signal tracks on the other side. The positive side of the 5-V supply can be laid out on the top side along with the circuit tracks for the components.

Otherwise if utilizing the older through-hole board technology, the PC board would be double-sided with the ground plane fitted in every space not occupied by circuit tracks or components. In fact, using multilayer board is very helpful here. In either case, the ground plane should be in three segments—one for the transmitter section, one for the receiver, and one for the audio section. Each of these three sections should be connected by a single ground conductor to the single-point ground (at one point only). The three sections should be separated by a Faraday shield (or the laminated capacitive "fence" referred to earlier) and grounded every half inch or less as discussed earlier. It is even wiser to entirely enclose the transmitting section with a box, which can be covered by a lid that snaps on to make good contact.

The ground plane serves multiple purposes. There are certain components on a transmitter that transfer energy at high frequencies. (These components in an RF transmitter would be considered "hot"; that is, they can radiate RF energy.) The ground plane provides isolation of "hot" component ground from those that are in a low-signal-level portion of the receiver. The ground plane is now a "zero-signal" base that affords a nonradiating surface and provides a short low-resistance (and AC impedance) path to the main signal ground.

Good bypassing techniques throughout are essential. This means that a separate bypass capacitor should be provided at every IC s power lead—speaking of which, I once had an engineer working for me who insisted on building prototype digital-circuit boards with no bypass capacitors at all. He insisted that the board had to work without any bypassing, so that when he added them he would be 100% assured that it would work without ripple and noise problems. (??) Theoretically, this might be a good idea, but it is very doubtful that it would ever work 100% without glitches, even if they were very minor and easily ignored. Even if they did, the random noise spikes will drive you up a wall chasing "ghost" pulses that intermittently show up and are generated simply by noise that goes just above the signal threshold of the logic or circuitry.

But back to the PC-board layout. As seen in Figure 11.2, the power, ground, and I/O section is grouped at the "bottom" of the board, near the bus connector. Here the bypass caps and ferrite beads we spoke of earlier are located, as well as the I/O driver devices.

Just above this, to the left, is the receiver section, with its own power and ground feeders. The local oscillator (if used) is shielded from the rest of the board by both a Faraday shield and a "can" with a snap-on lid that is spot-soldered when all testing has been completed. All circuitry in the receiver that runs at any frequency higher than about 20 Hz is very carefully laid out, with short leads and straight, short grounds.

Next, to the right is the audio section, which generally operates at frequencies in the 40-Hz–20-kHz range. This section *must* have good bypass techniques applied, and separate power and ground leads, with their own ferrite beads and capacitors connected between the feed to the audio circuitry and the bus connector.

To the right and below, is the worst of the bunch—the

transmitter section. Why is this the "worst of the bunch"? Because this section will have fairly high frequencies and draw considerable power. These two considerations dictate large bypass capacitors, large and straight power leads, a large ground plane, etc. It really should have a "can" or "box" surrounding it, with a snap-on lid that is well grounded. The only holes in this can should be those absolutely required for adjustments of pots, etc.

Grounding is of utmost importance here. RF and other spurious high-frequency signals require short, straight paths to ground to prevent radiation. And the ground plane for the entire transmitter portion of the board must be a single-point "drain." The power feeds to this section must be individually filtered.

If this section contains drivers or output circuitry that circulates fairly heavy currents, the feed from power must be larger than required to carry this amount of current— hopefully by 20% or more. The supply voltage must be taken into consideration . . . complete with any and all modulation at the outermost extremes of its envelope. This is because the supply voltage, multiplied by the worst-case current, must be allowed for in both the power supply requirements and the power dissipation characteristics.

Any connections to input/output ports must be made by way of a shielded path. Further, there should not be a single input path that parallels an output path (or vice-versa) without a ground between. Be sure that any two connector pins that carry *any* signals have a ground pin in between. And connect all shields at only one end—the originating end. By that I mean any shield ground should be made at the end from which that signal originates. No shields should be grounded at the receiving end. This will rule out any chance of ground loops from this cause.

The transmitter portion of the board is usually the highest-power portion of the board, and is definitely the most likely to radiate RFI. If there is a ground loop or other feedback path, stray capacity between circuits or inductive coupling is sure to cause problems. "Stray" feedback paths can cause oscillations in the receiver or transmitter where they are not wanted. Make sure everything is "bolted down" to a good hard ground.

These rules should give you a fair idea of the stringent requirements for circuit separation and filtering that may be needed for this type of composite-signal and integrated-function packaging. Whether the layout for your next project is done in the drafting room by a draftsman or on a computer by CAD, be sure you check the entire layout with the things we have pointed out here in mind.

In this chapter we have not given you a hard-and-fast set of rules to follow, spelled out in great detail—on purpose. Instead, what we have tried to do in this book is provide you with a very complete and comprehensive background of information to draw from, and let the insights you have acquired do the rest.

We sincerely hope we have done our job.

Good luck with your endeavors.

CHAPTER
12

Overall System Design

As this is the last chapter of this book, we shall take this opportunity to summarize all the design techniques we've discussed throughout the entire book. These and the other techniques given here will help to prevent both radiated and conducted EMI/RFI problems arising from either susceptibility or emission . . . but only if they are implemented properly.

We are going to start by looking at the most logical place to begin . . . where the system interfaces with the facility—the power cord and receptacle.

AC Power Input Filters

As pointed out early in this book, any AC power brought into the computer should be prefiltered (*for both directions*) before being fed to the power supply. This filter should be designed for both directions . . . from wall socket to computer *and* from the computer to the AC power line. The filter must be designed to match the impedances that it interfaces to . . . at both ends—input and output. It must be implemented in such a way as to prevent EMI/RFI being propagated in either direction—in or out. If designed with the ideas in mind that have been shown in this book, it need not be bulky.

An off-the-shelf power-entry module may be used in front of

and in conjunction with this filter. In fact, a power-entry module has many advantages—it can provide mounting convenience, extra filtering, and integration of required components such as a line switch and a fuse. The power-entry module can provide even more convenience if the one selected incorporates a line-voltage selector switch, as discussed earlier in this book. These input filters are manufactured by quite a surprising number of companies, and each type provides slightly different advantages.

Power Supplies

After the power cord and power entry to the enclosure, the power supply is the next major component in most of the smaller computers. Looking more closely at the available choices of power supplies, we find that there are several different types of supplies that we can choose from. Selection of the type of power supply that will be used is a very important and far-reaching decision. To help you make this important decision, we will take a closer look at the choices available. The following list of *types* of power supply must be considered very carefully—especially the relative advantages and disadvantages of each type. These must be weighed with the specifications and requirements of the particular system we are dealing with in mind.

Therefore, while reading the following discussion of the characteristics of each power supply type, bear in mind the particular requirements of the system you are working on. And while each of the following types of power supplies have relative good and bad points, there is also a range of *parameter* variations that must be considered. The system's input and output voltages and currents must be designed for, while at the same time fulfilling filtering and voltage regulation requirements dictated by the system specifications. At the same time, the overall power supply maximum physical size requirements must be considered, and will be a factor that will definitely affect the final decision.

Power Supply Types

Generally, you could say there are three main types of power supplies—the series-regulator or linear type, the square-wave switching regulator type, and finally the sine-wave switching regulator type. These are discussed more fully in the following sections.

Series-Regulator or Linear Power Supplies

Linear or series-regulator-type power supplies used to be the "old standby." They drew current in a fairly steady way that did not cause the power factor problems that are associated with some of the other types we have today. This was due to several factors. For instance, in a series-regulator power supply, the first device (besides the fuse and power switch) that the AC line voltage encountered was a power transformer. It acted as a choke-input, which meant the power supply could be designed for fairly low inrush current at turn-on. The series-regulator or linear power supply had several other advantages, such as clean, low-ripple output, with good voltage regulation under changing loads.

But series-regulator linear power supplies in general also have several disadvantages. Even though they are probably the least "noisy" type of supply, they are inherently bulky (due to the power transformer), and waste a good deal of energy—which is given off at the series regulator in the form of heat. They therefore cause heat dissipation problems. This is because the raw, rectified-but-"ripply" (lightly filtered) output voltage from the power transformer is fed to the regulator.

The input voltage to the series regulator is normally several volts higher than the output voltage required from the regulator. This is so that the series regulator can "smooth" it to a level below the lowest ripple "valleys"—although it dissipates considerable power in the process (especially if the regulator had to provide much current). Because the voltage dropped across the regulator must be multiplied by the current that the regulator is handling, the total power the regulator can handle is determined simply by the amount of heat that can or should be dissipated.

Another disadvantage of the linear regulator is that the power step-down transformer operates at AC line frequency (usually 60 Hz), and thus is (by design) large and bulky (read *heavy*). Recent advances in this technology have made this disadvantage less of a problem, however.

Switching Power Supplies

Advantages to be gained from a "switcher"-type power supply are far more numerous than those of the linear-regulator type. Among many others, one of the biggest advantages is the reduction in bulk and weight that can be realized from use of a switcher-type power supply versus a linear-regulator type. This comes about due to the higher frequencies at which power is handled in a switcher. But the use of square-wave switcher-type supplies can cause a few problems for the designer, too.

Square-Wave Switchers

Because square-wave switching power supplies draw current in great "gulps," they cause poor power-factor effects—as we have seen in prior chapters. The trend for some time was toward PWM (pulse-width-modulated) square-wave switchers, which switch under conditions of maximum current flow and therefore high di/dt waveforms—which causes a good deal of EMI and RFI. For these reasons, up until lately, switchers have had the reputation of being very "noisy."

Sine-Wave Switchers

Lately however, there has been a movement toward using *sine-wave switchers*, where the switching occurs at the zero current point in a sinusoidal input voltage. Think about it— since switching takes place at the point where the current is zero, practically no dissipation of power occurs here. Also, since we are dealing with a sine wave with low di/dt instead of a square wave, a lot less EMI/RFI is generated by this type of power supply.

Ripple levels are inherently low with a sine-wave switcher, due to the ease of filtering at the higher frequencies. Sine-wave voltages supplied to the power supply's rectifiers allow much quieter AC to DC conversion. This reduces overall EMI/RFI levels even more. Another big advantage of this type is, of course, the reduction in bulk, which as said earlier is a decided advantage when trying to fit the proverbial ten pounds of power supply into a five-pound portable computer.

Sine-wave switchers are not as plentiful as the square-wave variety . . . and, indeed, may not be for some time. The wide selection of parameters that is now available in square-wave switchers may never be matched by the sine-wave switching type. Only time will tell.

To continue in our quest for a "quieter" system, the next place to consider preventing emissions (and therefore susceptibility) is probably in the card cage/mother board combination (if a card cage or mother board is used).

Card Cage and/or Mother Board

Some computers will, of course, be built on a single mother board with no "back plane," and may or may not contain "slots" for extra add-on boards. Others may consist of many individual boards combined into a card cage with a back plane to implement the system—and probably will, in fact, combine several "CPU" (central processing unit) boards to allow parallel processing.

Other computers may be simply "embedded" processors which, along with certain peripheral devices, drive an electromechanical (or possibly a pneumatically or hydraulically driven) machine of some sort. Some of these embedded-processor systems may perform as mundane a purpose as controlling an oven, stove, dishwasher, or other appliance.

In any case, whatever the application, filtering and/or shielding must be provided to prevent EMI/RFI generated in the computer from escaping to the outside world (and conversely, outside interference sources from disrupting the operation of the computer).

As we discovered previously, there are hundreds of companies who specialize in equipment for this purpose. I suggest that you collect information from the many manufacturers on as many types of equipment designed for prevention of EMI and RFI as you possibly can. And weigh the advantages and disadvantages of each for yourself by evaluation testing.

But to continue in the same vein as the foregoing discussion, the next part of the computer that we need to consider is enclosures.

Enclosures

The enclosure that encompasses the computer must be radiation-proof—in both directions. That is, it must contain any RFI or EMI energy within—preventing its escape, but it must also shield the computer from any external interference. To do this well, the enclosure must be designed with this in mind . . . from the very beginning.

Then, the very important and crucial "make or buy" decision must be made. And in either case, the enclosure must have been designed and built to prevent emissions in *either* direction, as we have discussed many times, earlier in this book. Again, I recommend that you accumulate as many catalogs from enclosure manufacturers as possible, and equip yourself as well as possible to help make this inevitable and important decision.

Cabling

Inevitably—the link between the various boards within a cardcage, between individual boards and the outside world, and between the system processor and its peripherals must be some type of cable. Bear in mind that this cable is *the most important single potential cause* of RFI and EMI.

But to be perfectly honest, the subject of cable selection is actually too vast a topic to be completely covered here—with too many variables to cover in our limited space.

But we will go into some of the more important considerations that will have to be dealt with. Because cables used in a system must not only meet tough requirements for form and fit, they must also be selected using other important criteria—such as shielding requirements (if any), number of cables versus space available, size of the individual conductors in each cable, number of conductors required per cable, and many other considerations.

Take the case of a multiple-board system complete with a back plane. Even though the computer will have an internal back plane, there may still possibly be requirements for some type of cables to connect the boards to each other—or the boards to a front panel, or on the back panel, where the peripherals plug in. These cables will play a very big part in the overall scheme for the suppression of noise going into or out of the computer.

Also, frequencies may play a big part in the selection of cable types to be used. Waveshapes (square-wave, sinusoidal, etc.) may require coaxial cables . . . possibly with such special characteristics as double shielding or two conductors within one shield. Or a special requirement may demand that the conductors be twisted, in order to prevent cross-talk.[1]

So it goes . . . and now on with the quest . . .

Peripherals

Any *peripherals* included must be designed with EMI/RFI susceptibility and emission in mind, of both the radiated and conducted varieties. Input/output (I/O) cables connecting them (if any) must be properly shielded, with the shields connected to protective ground at one end only. All grounds must meet at a single point. Earth or "green-wire" ground must not be used as system or signal ground. (More about this under "Grounding and Power").

Drivers and receivers must be tailored with the utmost care, setting them by the proper trimming resistors to the expected

---

1. The transfer of a signal from one conductor to a conductor running parallel—usually through capacitive effects between the conductors.

I/O cable requirements. If EIA RS422 is chosen as the I/O specification, the tailoring is very straightforward and fairly simple.

Grounding and Power

The matter of grounding, especially when considering the entire system—is not trivial, as seen in earlier chapters. It requires much in the way of insight into many varied factors and conditions, which are difficult to predict, at best. It requires forethought on such unlikely things as weather, possible locations the machine may be used in, altitude considerations, and many others—as seen in foregoing chapters. For the most part, we can simply say that grounding must be done in such a way as to prevent ground loops and cross-talk, as well as noise on the signal ground.

Power input is always a problem, but with the "ounce of prevention" we have discussed in this book, which can be built into our product, there will be much less likelihood of the requirement for "a pound of cure."

If you happen to be at the user or field-support end of the business, be sure that the installation team has done a good job of specifying the environmental or site preparation that is required for installation. AC power noise levels, grounding requirements, and other considerations are equally important. Then be sure that the people who did the site preparation followed these requirements.

Other Considerations

In order to apply and implement the things we've learned in this book, we need to have a very good grasp of the volume and scope of products available for this purpose. Obviously, there is not enough room in this book to list each and every piece of test equipment and device designed for use in the prevention of ESD, RFI, and EMI.

The best we can hope for is a comprehensive list of the *types and categories* of these products—with a less-than-complete, but representative list of the manufacturers who make them. It is with this in mind that I shall include herein just such a listing of the types and categories of dedicated EMI/RFI products available on the market at this writing . . . besides those components already covered earlier in this book.

A listing of these categories of products follows.

Available Products

The products available for inclusion in systems or for use in testing them span several different categories, beginning with the very smallest components to the overall enclosures and external test equipment. The list we shall compile will cover such things, organized from the outside in, beginning with the test equipment. They are organized by categories, such as:

- Test equipment and systems
- Enclosures: cabinets, card cages, etc.
- Filters, filter modules, etc.
- Power supplies
- Shielding materials
- Toroids, inductors, etc.

Test Equipment and Systems

As explained earlier, a complete and totally encompassing list of test equipment and systems available "off-the-shelf" for use in combating EMI and RFI would be prohibitively large[2] and time-consuming for inclusion in this book. For this reason, included here is simply a short but fairly comprehensive list of the different types and categories of test equipment engineers, can avail themselves of. For instance, there are:

- Waveform generators that emulate the EIA–established-as-standard waves that have been developed as a model for AC power surges and spikes

---

2. A list such as this would exceed the length of this book.

- High-voltage ESD discharge generators, which provide repeatable, controlled discharges for use in ESD testing
- Oscilloscopes designed specifically for capturing, tracking, displaying, and characterizing voltage or current waveforms that cause various problems
- Strip-chart recorders, used for recording real-time waveforms during a specific period of time or section of a test
- Meters capable of measuring all sorts of phenomena, such as electrostatic voltages, very-short-duration current spikes, and magnetic fields and/or RF fields
- Power-line monitors used to capture and record spikes, surges, and changes in frequency or voltage of AC power lines
- Radiated EMI/RFI measurement devices, utilizing antennas and measuring equipment
- Conducted EMI/RFI measurement equipment
- Test equipment that generate either radiated or conducted EMI and RFI, as well as ESD.

As I said above, a complete list of all the equipment available for combating EMI, RFI, and ESD would more than fill a volume the size of this book. A complete list of manufacturers who build equipment for the purpose is equally impressive, and becoming more so every day.

But a few of the more popular EMI, RFI, and ESD test equipment and device manufacturers would include the old standbys such as:

Boonton	Frequency and RF voltmeters, etc.
B & K	From meters to entire test systems
Fluke	Meters, generators, etc.
HP (Hewlett-Packard)	The gamut, from 'scopes to generators, meters, and systems
Tektronix	Oscilloscopes and other test equipment
Nicolet	Digital Testing Equipment

This list would also have to include:

Advanced Protection Technologies
Best Products
Charleswater Products
Elgar Corp.
Fair-Rite Products
Gould Electronics
Key Tek
Saft
Topaz

and many, many others too numerous to list.[3]

Cabinet Enclosures

Cabinets that are specifically designed for RFI/EMI emission (and susceptibility) prevention are sometimes also targeted for TEMPEST applications. They may also be designed to fulfill MIL-SPEC requirements. A line of these enclosures are available from several different manufacturers, to fit most standard size requirements.

They are available tested to MIL-STD-285 for shielding effectiveness, from several manufacturers, "off the shelf." The enclosures are available in the standard 19- or 24-in. panel widths, with several frame depths. (The 30- and 36-in. depths are the most popular.) Panel-opening heights are offered in many standard sizes, which makes selection easy. These enclosures (usually known as racks) feature such options as removable side panels, hinged front and back doors, removable floor panels, fan and blower grilles and shrouds, and casters.

Again, unfortunately, a list of manufacturers of these products would be too large to include here. Instead, the reader is advised to consult the local distributor or mail-in "bingo cards" found in such publications as *EDN*, *EE*, and *Electronic Design*, and others with the ad number circled for more information from the manufacturers.

3. We apologize if you or your product are not listed here. Information was requested from those companies whose advertisements appear in the magazines serving the industry, but may not have been received in time.

Card Cages and Enclosures

Most card cages and enclosures are tailored to a particular line of products or a type of data-bus configuration. They can, however, be found with shielding, grounding, etc. arranged for meeting EMI/RFI emission and/or susceptibility requirements. These, too, come in such a variety of configurations that to include a list would make this chapter prohibitively large.

Just remember to follow the rules we have established throughout this book in selecting these products.

Filters and AC input modules are available in an almost infinite variety, and from many different manufacturers. Filters are available from the tiny ferrite devices used in printed-circuit boards to the huge AC power filters used in input power lines. As you remember, equipment from one particular filter manufacturer was described in Chapter 5, complete with pictures.

Shielding, Materials, etc.

The word "shielding" covers several types of materials used for several different types of interference. Mu-Metal, copper, and other shielding materials are available for use in shielding those portions of a system that require it. For instance:

- Mu-metal is used against magnetic fields to prevent them from emanating from an electromagnetic device beyond the limits of the shield itself. Mu-metal is a very ductile, bright metal material—typically used for magnetic interference shielding, for prevention of RFI and EMI susceptibility *and* radiation or emissions.
- Copper is often used between windings in power transformers as a Faraday shield against circulating electrical currents in transformers, cross-talk, electromagnetically induced eddy currents in nearby circuits, etc. On the other hand, it can also be used for sealing doors and other movable portions of an enclosure, and many other applications where a highly conductive, ductile material is required to shield parts or areas electrostatically.

- Beryllium–copper alloy spring-gaskets are used in joining doors and other movable panels for electrical connections for conductivity. Beryllium–copper alloy is best used where the material needs to be very conductive, but springy as well.
- Iron or ferrous metal shields are used where the magnetic field can be shunted through the ferrous material without its being propagated beyond the shield limits.
- Conductive gasket material is available both formed and unformed for use in EMI/RFI seals on doors, fan enclosures, and hundreds of other applications.

Toroids and Inductors

One of the most important devices with advantages that may not be inherently obvious to the most casual observer might be such devices as toroid cores and inductors. These devices are particularly useful in both the prevention and cure of radiated EMI and RFI, since the fields are for the most part *completely contained within the core material*. These toroids cores are used in transformers for power supplies, for inductors in input and output filtering devices, and even in I/O lines.

Back to Basics

To change the subject somewhat (yet to one that is related to our discussion here), do not be misled by those who would have you believe that cost is more important than reliability. While it is true that you can sell software that doesn't exist (known as "vaporware"), it has been proved over and over that you cannot *continue* to sell hardware that does not work.

The "rules of thumb" that should be applied in hardware manufacture might include the following:

1. At the board level:
   a. Good single-point grounding techniques should be applied.
   b. Power input should be filtered both by bypass capaci-

tors of two different values to cover the entire "threat band" of frequencies, in combination with use of ferrite beads.
   c. Plenty of bypass capacitors should be used at the power pins of ICs on the board.[4]
   d. The ground plane should extend to every place on the board not occupied by traces or insulating gaps between traces.
   e. There should be no unterminated inputs to ICs.
2. On back planes
   a. Plenty of bypass on DC power tracks
   b. Good single-point ground techniques
   c. No unterminated lines
   d. Good ground-plane techniques
   e. For power and grounds, large wire size that is as short as possible
3. Power supplies:
   a. Good AC power-entry EMI/RFI filtering
   b. Short leads everywhere
   c. Good EMI/RFI filtering on outputs
   d. Good shielding techniques throughout, following the rules in this book
4. Good shielding and terminations on I/O lines and cables, with lengths held to less than maximum length specifications
5. Good peripheral design throughout, including all the above

Going against the rules of thumb outlined in this book not only allows, . . . but begs for . . . problems due to EMI, RFI, and ESD.

And remember—if you get into trouble with ESD, RFI, or EMI problems, find yourself a good consultant. They might be a little expensive in the short term, but they can make the difference between a product that will fail FCC tests, resulting in the possibility of the product never making it to market (or worse,

---

4. I once knew an engineer who had a peculiar outlook on this (see Chapter 11).

being pulled off the market after it hits the streets). Or susceptibility may cause the product to fail in the field.

In the words of "Mr. Goodesign," the good engineer, stick to good design techniques, and help us to "stamp out ESD, RFI, and EMI!"

THE END

────────── APPENDIX ──────────

Other Sources of Information on This Subject

I once had the extreme pleasure of meeting a person whom I personally and profesionally admire—Mr. J. Fred Kalbach. I feel that he is "Mr. Grounding Expertise personified." Mr. Kalbach is a consultant and a recognized authority on noise-suppression techniques. He is the author of an article that appeared in the January 1982 issue of *Digital Design Magazine,* entitled "SYSTEMS Designer's Guide to Noise Suppression."

In this article, he expounded on the inevitable results of failure to follow the proper noise-suppression and grounding rules. EMI requirements are discussed as related to FCC requirements. A very informative passage on common-mode versus normal-mode noise voltages appears in this article, as well.

Mr. Kalbach is also the coauthor of the renowned government publication entitled "FIPS PUB 94." This publication is a very authoritative source of power and ground-noise-suppression techniques in the electrical environmental considerations of computer installations.

Other very authoritative sources for information on this and related subjects are such publications as follows.

1. The IEEE Publication, *IEEE Guide for Surge Voltages in Low-Voltage AC Power Circuits.* IEEE Designation: IEEE STD 587-1980.

2. James R. Huntsman and Donald M. Yenni, Jr., "Test Methods for Static Control Products," Static Control Systems, 3M Company.
3. J. F. Kalbach, "Designer's Guide to Noise Suppression," *Digital Design Magazine* (January 1982).
4. Application Notes 106 and 111, by KeyTek Instrument Corp.
5. Engineering Notes entitled "Ferrite Components as EMI Suppressors" by Fair-Rite Products Corp.
6. A bulletin by John Howard of Radiation Technology, Morgan Hill, California.
7. A booklet entitled *Surge Protection*, written by Ron Chapman, Vice President of the publisher—Advanced Protection Technologies, Clearwater, Florida.

Glossary

The following is a list of definitions for terms used in this book.

Ambient temperature Temperatures found inside the enclosure of a particular portion of a system, and which is in direct proximity with the components of a printed circuit board or part.

ATE (Automated Test Equipment) Equipment that is automated (usually computerized) to test a product automatically at very high repetition rates.

Bingo card Information-retrieval cards contained in various magazines, containing a list of numbers that correspond to ads throughout the magazine.

Breakdown voltage The voltage value at which a semiinsulating medium suffers a calamitous current surge due to ionization of that medium.

Buffered outputs IC outputs that are internally isolated from the portion of the IC that is the "flip-flop" element itself, so that a forced change at the output temporarily will not force a permanent change of state.

Burn-in The process of "cooking" new parts at elevated operating temperatures with normal or increased operating voltages applied for the length of time required to assure they will continue to work without premature failure due to "infant mortatlity."

Common-mode noise Noise signals that are present on

both sides of the AC power line (hot and neutral) as referenced to ground.

Corona Discharge The tendency of a high AC voltage at fairly high frequencies to form a "corona" of highly ionized gas around the most pointed part of the circuit.

Cross-talk The transfer of a signal from one conductor to a conductor running parallel—usually through capacitive effects between the conductors.

Daisy-chain A series of connections, each following one after another, through a single cable with multiple stops.

Diathermy machine A large heavy machine that generates high-power 27 MHz RF, to be radiated from a coil that could be passed over a patient's body in the manner of a metal-locator. The high-power RF then induces artificial fever in that part of the body, to aid in healing.

Differential-mode noise (also known as transverse-mode noise)—Noise signals that appear on one AC line as referenced to the other (hot-to-neutral)

EMI (electromagnetic interference) EMI is a "nickname" for the class of interference caused by magnetic fields—either directly through the air or by the magnetic lines of force cutting a conductor, causing a resultant unwanted signal in that conductor.

EMI emissions EMI that is emitted from a computer or similar device, which can consist of either of two varieties:

   a. **Radiated** Electromagnetic interference emitted from a device and carried through air or a similar medium.
   b. **Conducted** Electromagnetic interference that is conducted via metallic objects (wires, pipes, etc.)

EMI susceptibility The inability of a computer to withstand electromagnetic interference of either of two varieties:

   a. **Radiated EMI susceptibility** Electromagnetic interference from a source external to the device, carried through air or a similar medium.
   b. **Conducted EMI susceptibility** Inability of a computer or peripheral to withstand electromagnetic interference,

which is conducted via metallic objects (wires, pipes, etc.).

EO (engineering order) Sometimes called "engineering change order" or "engineering change notice" (ECO or ECN).

ESD (electrostatic discharge) ESD is defined as the high voltage "spark" caused by a "static charge," such as results when you walk across a rug and then touch a metal object.

Flashover The point at which a material (normally nonconductor) has ionized until its impedance is low enough to allow current flow at the voltage being applied.

Flip-flop A device that accepts up to two inputs plus a clock pulse and/or "set" and "reset" inputs, and which at the clock input causes the output to assume a certain state.

Glitch A voltage variation, usually a result of a disturbance caused by some other device. This voltage variation may be "induced" into the circuit by EMI, or "conducted" from somewhere else. The glitch is of very short duration (usually less than a few microseconds).

Ground plane A large, flat metal plate (or plating in the case of a printed-circuit board) that provides a convenient connection to the system signal ground and also provides shielding and capcitive-coupling for parts. This term is also applied to the background sheet metal or earth used as a base for a radiating antenna.

Ground voltage gradient The progressive voltage drop through earth ground caused by current flow through a poor conductor (the earth itself), which increases with the amount of current flowing through that ground.

Infant mortality Defined as failure during the first 72 hours of operation.

Insertion loss The ratio of power contained in a signal at the input to a device versus the power remaining at the output, expressed in decibels.

Noise A term loosely used to describe *any* extraneous signals that appear where they do not belong. The title of "Noise" may even be loosely applied to one of the above *causes* of a ZZAAP!

OEM (original equipment manufacturer) Not (as you

might think) the manufacturer of the original equipment, but instead the manufacturer who incorporates another manufacturer's equipment into his or her design or product.

Overshoot The property of a waveshape having a steep wavefront that continues rising past the point it should, and then oscillates slightly around the maximum DC voltage level, then damping out after a few cycles.

Passband The band of frequencies in a filter that exhibit the lowest insertion impedance to a signal being propagated through that filter.

Punch-through A term applied to overvoltage (usually also reverse-voltage) effects that cause the internal insulating layers in a semiconductor to break down and allow current flow, thereby rupturing that insulation irreparably and causing a short circuit at that point.

Quiet room A room totally shielded and sealed against radiation either leaving or entering the room. It also may have radiation-absorbing material affixed to walls and ceiling to provide as near a perfect radiationless environment as possible for performing antenna and RF tests.

Read-only memory (ROM) A read-only, programmed integrated circuit that usually contains a "monitor" for the software operating system of the computer.

RFI (radio-frequency interference) RFI is defined here as interference caused by radio-frequency energy manifested where it does not belong. RFI can be carried by either air or by wire, or as a "stray" radio signal.

RFI susceptibility The inability of a computer to withstand onslaughts of RFI that come from elsewhere in its environment. This RFI can consist of one of two varieties:

    a. **Radiated RFI** RFI that is radiated through the air as radio signals
    b. **Conducted RFI** RFI that travels on wires (AC power wiring, I/O cables, interunit wiring, etc.).

RFI emissions Radio-frequency interference that comes from a device outward, in either of two varieties:

a. **Radiated RFI** RFI that is carried through air or similar media as radio signals.
b. **Conducted RFI emissions** RFI that is conducted through AC power or other wiring.

Ride-through The capability of a power supply to store enough input current to continue to supply output voltages at the proper level during temporary loss of input power.

Screen room A metal-framed room with metal screen covering (usually copper), with a metal door bonded to the rest of the room and equipped with EMI gaskets or similar protection to prevent any radiation, EMI, RFI, or other electrical noise from entering or leaving the room.

Surge A sudden increase in the input AC voltage.

Sag A corresponding decrease in input AC voltage.

Tesla coil An air-core transfomer with a very high ratio of secondary turns as compared to the primary, which when excited by a high-frequency RF primary source, boosts the output voltage to thousands of volts.

Thermal-intermittent Intermittent problems brought to the surface when a part is either raised or lowered in temperature.

Thermal shock The physical shock experienced by a component that has either been at operating temperature for some time and suddenly cools off; or vice-versa, is dead cold and is rapidly heated to oprating condition after being powered up.

Transient A transistor, asynchronous electrical impulse, that appears on a conductor and usually results in some undefined and unwelcome reaction in a computer.

Transverse-mode noise Noise that appears on one AC line as referenced to the other (also called "differential mode").

Tuned circuit As used here, a circuit path that displays an attenuation curve whose point of least attenuation occurs at some particular frequency (also known as a *band-pass curve*).

Wilmhurst machine A wheel, made from aluminum with a set of copper "fingers" at intervals around the wheel on both sides, with the copper and aluminum insulated from each other. This wheel was mounted on an axle, with bearings through a stand and a set of collecting brushes. The machine generates high voltage when rotated.

ZZAAP! For the purposes of this book, defined as an incident *resulting* from a "transient" (any of the above-defined noise sources).

Index

AC power, outside computer room, 127
AC power input filters, summary, 205–206
AC power variations
  drop-out, 45–47
  glitches, 41–42
    sags, 39–40
  line conditioners, 51–54
    lightning arrester, 53–54
    types, 51–53
  noise, 42–45
    common-mode, 43–44
    transverse mode, 43, 44–45
  power factor effects, 54–57
  spikes, 40–41
  surges, 34–39
    aftermath of surge, 37
    causes, 35–37
    lightning, 36–38
    voltage surges, effects, 38–39
  transient suppression devices, 48–51
    breakover diodes, 50–51
    gas tubes, 50
    MOVs, 49
    problems, 51
    zener dioxide, 49–50
    transients, 33–34
Air compressor, as cause of EMI, 13–14
Ambient temperature, definition, 223
  and ESD, 157–160
American National Standards Institute (ANSI),
  and ESD testing, 168
Application-specific integrated circuit (ASIC), 189
Auto repair shop, as computer facility, 133, 144
Automated Test Equipment (ATE), definition, 223
  and burn-in, 150, 152
Autotransformer, 44

Back planes, hardware, 218
Bingo card, definition, 223

Board-level hardware, 217–218
Boat, as computer facility, 133, 142–143
Breadboarding, 185–191
Breakdown voltage, definition, 223
Breakover diodes, 50–51
Buffered outputs, definition, 223
Burn-in
  automated test equipment (ATE), 150, 152
  definition, 147, 223
  and infant mortality period, 147–148
  scenario without testability, 151–152
  TDBI tests, 149, 150–151, 152–153
    case, 153–154
  testability, built-in, 154–155
Bypassing, and circuit board design, 197
  good techniques, 201

Cabling, summary, 210–211
Cabinet enclosures, 215
CAD/CAM, *see* Computer-assisted design/manufacturing techniques
Capacitive-input current waveshape, 55–56
Capacitive-input switching power supply, 56
Card cage and/or motherboard, summary, 209–210
Card cages and enclosures, 216
Cathode-ray tube (CRT), and zzaap, 16, and problem description of error, 19–20
Central processing unit (CPU), of computer, 15
Circuits
  segregation, and circuit board design, 197–198
  types, segregation, 199–203
Common-mode noise
  definition, 223–224
  and power variations, 43–44
Complementary metal oxide semiconductors, 15–16
Compliance Engineering, as source of information, 118
Component surface temperatures, 158
Computer-assisted design/manufacturing (CAD/CAM) techniques, 185–186
Computer crashes, reasons
  electrical noise, effects, 15–16

electromagnetic interference (EMI), 11–15
  case, 12–15
electrostatic discharge (ESD), 8–10
errors, common, 18–25
  other factors, 24–25
  problem description, 19–20
  repair, 21–22
ground-loop phenomenon, 25–26
ground noise, 26–31
  every-other-wire-a-ground, 29–30
  requirements for low noise, 28–29
noise, designing, 17–18
radio-frequency interference (RFI), 10–11
zzaap, indications, 16–17
Conductive materials, for use in preventing RFI, 117
Cooling air supply temperature, 158
Corona discharge, definition, 224
Crib death syndrome, *see* Infant mortality
Cross-talk, definition, 224

Daisy chain, definition, 224
DC power supply to circuit board, 196
Dedicated circuit, 130
Dedicated ground traces, 196
Dedicated isolated (third wire) ground, 130
Desert conditions and ESD, 161
"Designer's Guide to Noise Suppression," 222
Devise under test (DUT), 170
Diathermy machine, 169
  definition, 224
Differential-mode noise, definition, 224
Digital Design Magazine, 220
Dranetz AC voltage monitor, 128
Drop-out, power, 45–47

EDN, as source of information, 118
EDN News, as source of information, 118
EE Evaluation Engineering, 77
  as source of information, 118
EIA RS422 signal transmission, 92
  RS232C versus RS422, 92–93
Electrical noise, effects on computer crashes, 15–16
Electrical stress, 174, *see also* Thermal shock

Index / 231

Electromagnetic interference (EMI), definition, 4, 224
Electromagnetic interference (EMI), and computer crashes, 11–15
  case, 12–15
Electronic Design, as source of information, 118
Electronic Engineering Times, as source of information, 118
Electronic Products, 118
Electronic switching regulators, 52
Electronic voltage regulators, 52
Electronics, as source of information, 118
Electrostatic discharge (ESD), and computer crashes, 8–10
Electrostatic discharge (ESD), definition, 4, 225
Electrostatic discharge (ESD), and environmental effects
  ambient temperature, 157
  cooling air supply temperature, 158
  enclosure air supply temperature, 158–160
  humidity, effects, 160–168
    hypothetical requirements
      design specifications, 162–163
      EMI/RFI requirements, 163–164
      temperature and humidity/pressure requirements, 164–165
      voltage and current requirements, 165
    Murphy's Law, 165–168
  testing, 168–172
Embedded processor design, 191–193
EMI, example of 97–102, *see also* Lightning, 99–101
  shielding, steps necessary, 101–102
EMI emissions, definition, 224
  conducted, 224
  radiated, 224
EMI emissions tests, 74
  conducted, 74, 78
  purpose, 78–79
  radiated, 74, 78
EMI filter equivalent circuit, 71–72
EMI susceptibility, definition, 224
  conducted, 224–225
  radiated, 224

EMI susceptibility tests, 74
  conducted, 74, 77–78
  purpose, 78–79
  radiated, 74, 77
Emission, noise, *see* Noise susceptibility and emission
Enclosure air supply temperature, 158–160
Enclosures, summary, 210
Engineering order (EO), definition, 19, 225
  repair problem of computer crash, 21–22
Environmental aspects of installing computer, 30
Environmental effects and ESD, *see* Electrostatic discharge and environmental effects
Equipment, test, 212–215
Error messages, 17
Errors causing computer crashes, common, 18–25
  other factors, 24–25
  problem description, 19–20
  repair, 21–22
  result of factors, 23–24
ESD, 96–97, 98–99, *see also* Electrostatic discharge; Lightning
European Computer Manufacturers Association (ECMA), and ESD testing, 168
Every-other-wire-a-ground rule, 29–30
External-to-system grounds, 94–96

FAA/NASA Symposium on Lightning Technology, applications note, 47, 106
Facility environment, and computer system problems, 127
Faraday shield, 44, 117
Federal Communications Commission, requirements for low interference emissions, 17–18
"Ferrite Components as EMI Suppressors," 222
Ferroresonant transformer devices, 51–52
Field service bulletin, of list of equipment incompatible with computer systems, 110
Filter input, 67–69

Flashover, definition, 225
Flashover point, 9
Flip-flop, definition, 225

Gas Tubes, 50
Glitch, definition, 225
Glitches, power, 41–42
Green-wire ground path, 25
Ground-loop phenomenon, and
  computer crashes, 25–26
Ground loops, breaking, 44–45
Ground noise, combating, RS232C
  versus RS422, 92–93
Ground noise, and computer crashes,
  26–31
  every-other-wire-a-ground, 29–30
  factors, 27
  requirements for low noise, 28–29
Ground plane and ground tracks
  definition, 225
  and printed circuit board design,
  196–197
Ground voltage gradient, definition, 225
Grounding
  and circuit board design, 202
  and power, 212
Grounding requirements, and lightning,
  see Lightning, and grounding
  requirements

Hardware manufacture, basics, 217–219
High altitude conditions, and ESD, 161
High instantaneous currents, 177
Humidity, effects in computer room
  environment, 160–168
  conditions, 161
  EMI/RFI requirements, 163–164
  hypothetical requirements
    design specifications, 162–163
    temperature and humidity/pressure
      requirements, 164–165
    voltage and current requirements,
      165
  Murphy's Law, 165–168

"IEE Guide for Surge Voltages in
  Low-Voltage AC Power Circuits,"
  221
Inductive choke input, 56, 57
Inductor input, 67

Inductors and toroids, 217
Industrial manufacturing plant, as
  computer facility, 133, 144–145
Infant mortality and burn-in, 147–148
  definition, 225
Information, sources, 221–22
Insertion loss, definition, 225
Installation site considerations for
  computers, 132–145
Instantaneous voltage changes at
  "turn-on," 177
Institute of Electrical and Electronics
  Engineers (IEEE)
  Symposium on Electromagnetic
    Compatibility, application note
    111, 47–48
  standard transient waveform, 16
Internal resistance, reduction, 178
International Electrotechnical
  Commission (IEC), and ESD testing,
  168
Invertor output waveshape, effects,
  139–140
Ionized air, 89, see also Lightning
Isolation transformer, 53

Lightning, as power surge, 36, 37–38
Lightning arrester, 53–54
Lightning, and grounding requirements
  ground potential disturbances, 81–86
  ground voltage gradient, 86–88
  phases, 87, 88–92
  voltage waveform of lightning strike,
    90, 91
Line conditioners, AC, types, 51–53
  electronic switching regulators, 52
  electronic voltage regulators, 52
  ferroresonant transformer devices,
    51–52
  isolation transformer, 53
  switching regulator, 52
  uninterrupted power supply (UPS),
    52–53
Locking compound, 20

Machine shop, as computer facility, 133,
  144
Magnetic field expansion, 12
Mechanical stress, 174, see also Thermal
  shock

Index / 233

Metal oxide varistors (MOVs), 46
  as surge suppressors, 60–61
  as transient suppressor device, 49
Mother board, summary, 209–210
Motorhome, for computer installation site, 133, 134–142
  conclusions, 142
  electrical system, 134–135
  invertor output waveshape, effects, 139–142
  uninterruptable power supply (UPS), alternative, 135–139
Murphy's Law
  versus controlled conditions, 121–130
  case, 122–123, 128–130
  factors, 123–127
  and protyping, 189

Natural resources in soil, as factor not within control, 126–127
Noise, definition, 4, 225
Noise, designing out, 17–18, *see also* Computer crashes
Noise, power, 42–45
  common-mode, 43–44
  transverse-mode, 43, 44–45
Noise filters and surge suppressors, *see* Surge suppressors and noise filters
Noise susceptibility and emission
  definitions, 103–106
  duality theory, 106–107
  paths, 107–109
    cases, 109
    EMI and RFI sources, 108–109
  other sources of information, 117–119
  prevention, components available, 109–117
    equipment, 111
      conductive material, 117
      power-entry modules, 111–113
      RFI/EMI enclosures, 116
      RFI/EMI filtering switching, 113–116
      shielding, 116–117
Noise susceptibility, low, design requirements, 28–29

One spot (single-point ground), 85
Original Equipment Manufacturer (OEM), definition, 23, 225–226
Overshoot, definition, 226

Passband, definition, 226
Paths of noise
  cases, 109
  EMI and RFI sources, 108–109
Peripheral design, good, 218
Peripherals, summary, 211–212
Physical stress, 174, *see also* Thermal shock
Power entry modules, as equipment for reducing or preventing RFI, 111–113
Power factor effects, and AC voltage, 54–57
  capacitive-input current waveshape, 55
  capacitive-input switching power supply, 55, 56
Power input device specifications, 66
  connections, 69–70
  filter input, 67–69
  inductor input, 67
  mechanics, 70–73
Power station, as computer facility, 133, 143–144
Power supplies, hardware, 218
Power variations, *see* AC power variations; Drop-out; Glitches; Sags; Spikes; Surges
Preventive design techniques
  embedded processor design, 191–193
  other preventive techniques, 193–194
  protyping and breadboarding, 185–191
    case, 186–187
    protype, 187–191
Printed-circuit-board design and layout
  design techniques, 195–198
    bypassing, 197
    circuit segregation, 197–198
    DC power supply to board, 196
    ground plane and tracks, 196–197
  layout techniques, 198–203
    segregation of circuit types, 199–203
Protyping, 185, 187–190
Punch-through, definition, 226
  and start-up failure, 177

Quality control and noise problem, 20
Quiet room, definition, 226

Radio-frequency interference (RFI), and
    computer crashes, 8, 9, 10–11
  variables involved, 10
Radio-frequency interference (RFI),
    definition, 4, 226
  conducted, 8
  radiated, 8
Rapid cooling, mechanical contraction,
    178
Read-only-memory (ROM), definition, 2,
    226
Reduced-instruction set computers
    (RISC), 189
Reliability factors, system controlled
    conditions versus Murphy's Law,
    121–130
  case, 122–123, 128–130
  factors, 123–127
  installation computer site
      considerations, 132–144
    auto repair shop, 144
    boat, 142–143
    industrial manufacturing plant,
        144–145
    machine shop, 144
    motorhome, 134–142
    power station, 143–144
    switchyard, 143–144
    welding shop, 143
  requirement specifications, 131–132
Resistive load current versus
    switcher, 65
Resonant or passband frequency, 84
Reverse breakdown, 178
Reverse-voltage knee principle, 49–50
RFI emissions, definition, 226–227
  conducted, 227
  radiated, 227
RFI emissions tests, 74, 75–77
  conducted, 74, 76
  purpose, 78–79
  radiated, 74, 76–77
RFI shielding, steps, 101–102
RFI susceptibility, definition, 226
  conducted, 226
  radiated, 226
RFI susceptibility tests, 74, 75–76
  conducted, 74
  purpose, 78–79
  radiated, 74, 75–76

RFI/EMI enclosures, as equipment for
    preventing or reducing RFI,
    116–117
RFI/EMI filtered switching power
    supplies, as equipment for reducing
    or preventing RFI, 113–116
Ride-through, definition, 227
RLC series circuit, loop equation, 83
RS422 cabling, use of for less noise, 28
  versus RS232C, 92–93

Sag, definition, 4, 227
Sags, power, 39–40
Saturable core reactors, *see*
    Ferroresonant transformer devices
Scientific American, 88
Screen room, definition, 227
Shielding, 216–217, 218
Soil composition, as factor not within
    control, 126
Space conditions and ESD, 161
Specifications, requirement, for
    installation of computer system,
    131–132
Spikes, power, 40–41
Square wave output, 139–140, 141
Start-up failures due to thermal stress,
    176–179
  diagnosing failures, 178–179
  mechanism, 177–178
Stress, thermal
  effects, 175–179
  types, 173–175
Strip-line, 190
Supplies, power
summary, 206
types, 207–209
  series-regulator or linear power
      supplies, 207–208
  sine-wave switchers, 208–209
  square-wave switchers, 208
  switching power supplies, 208
Surface-mount technology (SMT) form,
    188
Surge, definition, 4, 227
"Surge Protection," 222
Surge suppressors and noise filters
  equipment design and selection,
      60–73
    alternatives, 63–64

connections, 69–70
design techniques, 64–66
filter input, 67–69
inductor input, 67
inexpensive noise protection, 62–63
mechanics, 70–73
power input device specifications, 66
research, 73
research methodology, 73–79
  categories, 74–75
    conducted EMI emissions, 78
    conducted EMI susceptibility, 77
    conducted RFI emissions tests, 76
    conducted RFI susceptibility tests, 76
    EMI tests, 77
    radiated EMi emissions, 77
    radiated EMI susceptibility, 77
    radiated RFI emissions, 76–77
    radiated RFI susceptibility tests, 75–76
  RFI tests, 75
  purpose of tests, 78–79
  see also Transient suppressor devices
Surges, power, 34–39
  causes, 35–37
  lightning, 36, 37–38
  voltage surges, effects, 38–39
Switching regulator, 52
  electronic, 52
Switchyard, as computer facility, 133, 143–144
"SYSTEMS Designers Guide to Noise Suppression," 221

Temperature and humidity, pressure requirements, 164–165
Tesla coil, definition, 227
Test and measurement world, 47
  as source of information, 118
Test equipment and systems, 213–215
"Test Methods for Static Control Products," 222
Testability for burn-in
  built-in, 154–155
  scenario without, 151–152
  TDBI tests, 149, 150–151, 152–153
    case, 153–154

"Testing DC Power Supplies: Hidden Effects from the AC Power Source," 47
Theory of duality, 106–107
Thermal dissipation problems, 182, *see also* Thermal shock
Thermal shock
  controversy, 173–174
  finding failure modes, 179–181
  other effects, 183
  thermal dissipation problems, 182–183
  stress, 174–175
  stress, effects, 175–179
    diagnosing failures, 178–179
    intermittent, 175–176
    start-up failure mechanism, 177–178
    start-up failures, 176–177
Thermal stress, *see* Thermal shock
Thermal-intermittent, definition, 227
Toroids and inductors, 217
Transient, definition, 227
Transient suppression devices, 48–51
  breakover diodes, 50–51
  gas tubes, 50
  MOVs, 49
  problems, 51
  zener dioxide, 49–50
Transverse-mode noise,
  definition, 227
  and power variations, 43, 44–45
Truck or van, as hostile environment for small computers, 133
Tuned circuit, definition, 227

Uninterrupted power supply (UPS), 52–53
  an alternative, 135–139, *see also* Reliability factors

Value-added reseller (VAR), 18
Van, as hostile environment for small computers, 133
Voltage surges, effects, 38–39, 48
Voltage and current requirements, 165
Voltage waveform of lightning strike, 91

Weather, as factor not within control, 123, 124–126

environmental factors, other, 126
humidity, 125
seasonal temperature norms, 125–126
thunderstorm, 124
air pressure, 125
    ground conductivity, 125
    humidity changes, 124–125
Welding shop, as computer facility, 133, 143

Wilmhurst machine, 169
    definition, 227–228

Zapper, 169
Zener diodes, 49–50
Zzaap
    definition, 2, 3–4, 228
    indications, 16–17, *see also* Computer crashes